Praise for *Changin*

‘I wish this book had been available when my children were younger! It’s wise and informative yet wonderfully easy to read. It moves from the biblical to the practical, the local to the global, making the complex simple and succeeding in being both deeply challenging yet also life-affirming and hope-giving.’
Revd Dr Dave Bookless, director of theology, A Rocha International

‘Challenging, engaging and practical, this is a wonderful resource to journey through either individually or as a family. It provides all sorts of ideas of how you can make a difference, and solid reasoning for the urgency of doing so. Read it and live it!’
Simon Guillebaud, international director for Great Lakes Outreach

‘Jamie, Debbie and David have written a comprehensive and practical guide to help us navigate our way through the choices and complexities of living more lightly on the planet. I love the tips and links, the quotes, the clear and challenging Bible studies, and the deep dives into the science and theology. This is one of the best and most accessible books I have read on *why* Christians should care about climate change, and *what* we can do about it. This is a book to be read and shared with your family, friends and church.’
Caroline Pomeroy, director, Climate Stewards

‘I love the fact that this book is global, intergenerational and intersectional in its scope, while still remaining accessible and full of hope. It presents a robust, biblical case for the need for Christians to engage with the work of climate justice and offers small and large ways in which this can happen. Most importantly, it provides a platform to hear from and celebrate young people who are striving to make a difference.’
Lynne Norman, Methodist children, youth and family team

‘The Hawker family takes on climate change! Biblical insights drive them to take action on one of the most urgent issues of our time. This book is packed with hints and tips about how YOU can make a difference.’
Dr Martin Hodson, author of *A Christian Guide to Environmental Issues*

'So often we hear the question, "But what can we do, and what if it doesn't make a difference?" This easy-to-read book offers a guide to faith-filled, faithful action undertaken as part of our discipleship. Highly recommend.'
Rachel Mander, on behalf of Young Christian Climate Network

'This is a remarkable, timely book on a fundamental remit of all Christians. As it takes a multilevel approach, from initial ideas to complex changes, it is really accessible to people of all ages and experiences. The mixture of stories, Bible studies, statistics and next steps keeps it readable and relatable. As a missionary, minister and parent who has home-educated for years, I can see so many uses for this resource – personal study/challenge, family devotion, small group project, educational material – and I highly recommend it.'
Hannah Prosser, Assemblies of God GB mission team

'This book is a wonderfully rich, engaging resource for meaningfully exploring and practically responding to one of the most pressing issues of our time. It facilitates intergenerational reflection on a wide diversity of environmental concerns in ways that are informative and interesting, considered yet challenging, and bold while biblical. Jamie's (very helpful) tips at the end of each chapter have really motivated our family to up our game in terms of creation care engagement!'
Richard and Louisa Evans, All Nations Christian College, UK

'Weaving together biblical insight, creative and accessible suggestions for taking action, and stories sourced from around the globe, this is a rich resource for Christians of all ages wanting to engage with the climate crisis. Highly recommended.'
Jo Swinney, head of communications, A Rocha International

'The book challenges its readers to engage both spiritually and factually with one of the biggest challenges facing humanity in the 21st century. Its clever, interactive and logical style makes it attractive to all ages. I enjoyed that each chapter was divided into sections. This made it easier to read and also easier to refer back to as needed. I really enjoyed Jamie's tips, which are full of interesting facts and easy steps that anyone can take as they journey to make their lifestyle more environmentally friendly. A very enjoyable and informative read.'
Simone Formolo-Lockyer, Latin Link

'As well as this work being packed with practical suggestions on what to do about climate change, it is refreshing to find biblical reflection on the environment which focuses not on the classically obvious passages about stewardship and respect for creation, but on many Bible texts which one would not normally associate with our responsibility towards the created world, such as the story of Jonah and the parable of the good Samaritan. This leads to such challenging questions as 'Who is my neighbour in the climate emergency?' Written by a mother who has not driven for 20 years, a father known as the Cycling Psychologist and a son who wants to be a train driver, this book brings lessons from those who know what they are talking about.'
Mark Greenwood, BMS World Mission

'*Changing the Climate* is a comprehensive, challenging and highly practical book that uses the Bible well with latest scientific evidence stating the case for us all to be more engaged in addressing climate change. The chapters are easily accessible, offering practical tips of things we can do to embrace creation care, interspersed with contemporary global examples where people and nations are suffering because of the current crisis they are experiencing. With space for personal reflection throughout, this book will help individuals, groups and families learn, adapt and change their behaviours as together we tackle one of the key issues of our time.'
Phil, director, AWM-Pioneers

'An extremely thoughtful, well-written yet practical book. Accessible and useful for all ages of your family and your church. It relates having a Christian faith to taking action on climate change. If you are a Christian (or even if you are not!) and you'd like to know what you can actually do to help save our planet – then this is the book for you.'
Richard Jackson, youth leader

Changing the Climate

Applying the Bible in a climate emergency

Debbie, David and Jamie Hawker

The Bible Reading Fellowship
15 The Chambers, Vineyard
Abingdon OX14 3FE
brf.org.uk

The Bible Reading Fellowship (BRF) is a Registered Charity (233280)

ISBN 978 1 80039 022 5
First published 2021
10 9 8 7 6 5 4 3 2 1 0

Acknowledgements
Unless otherwise stated, scripture quotations are taken from The Holy Bible, New International Version (Anglicised edition) copyright © 1979, 1984, 2011 by Biblica. Used by permission of Hodder & Stoughton Publishers, an Hachette UK company. All rights reserved. 'NIV' is a registered trademark of Biblica. UK trademark number 1448790.

Scripture quotations marked MSG are taken from *The Message*, copyright © 1993, 1994, 1995, 1996, 2000, 2001, 2002 by Eugene H. Peterson. Used by permission of NavPress. All rights reserved. Represented by Tyndale House Publishers, Inc.

Scripture quotations marked NLT are taken from the Holy Bible, New Living Translation, copyright © 1996, 2004, 2007, 2013. Used by permission of Tyndale House Publishers, Inc., Carol Stream, Illinois 60188. All rights reserved.

Scripture quotations marked GNT are taken from the Good News Bible published by The Bible Societies/HarperCollins Publishers Ltd, UK © American Bible Society 1966, 1971, 1976, 1992, used with permission.

'Refugees', in B. Bilston, *You Took the Last Bus Home* (Unbound, 2016), p. 208. © Brian Bilston. Printed with kind permission.

Every effort has been made to trace and contact copyright owners for material used in this resource. We apologise for any inadvertent omissions or errors, and would ask those concerned to contact us so that full acknowledgement can be made in the future.

A catalogue record for this book is available from the British Library

Printed and bound in the UK by Zenith Media

Dedicated to Chris and Susanna Naylor,
who pioneered A Rocha Lebanon; and Miranda Harris,
co-founder of the conservation charity A Rocha.
These three, who tragically died in an accident in October 2019,
set a beautiful example of following Christ faithfully
in caring for his world.

Acknowledgements

Thank you to the charities featured in this book, for providing information and pictures and for the fantastic work that you do.

Special thanks to the young people who provided case studies. You are an inspiration.

We also thank everyone who has provided photographs to illustrate the book, and **whatplastic.co.uk** for permission to use their waste-reduction pyramid (Figure 1 on page 19).

We are grateful to Simon and Alena Stead, Steve Gannon, Geoff Pritchard, Philippe Ndabananiye, John Steley, Jane and Andrew Longley, Richard Parncutt, Simon and Kati Butt-Bethlendy, Paul Farrant, Tony Potts, Heather Lovell, Katherine Hawker and the Effective Environmentalism and Ethical Omnivore Movement Facebook groups for inspiration, comments, information and encouragement.

Finally, sincere thanks go to Olivia Warburton, Alison Beek, Daniele Och and the rest of the team at BRF for bringing this book to fruition. You have special gifts, and you make a dream team.

Contents

Preface

Planet earth is in crisis. Earth will survive. Humans, or at least our way of life, may not.

If we carry on as we are doing, we are on course for an average temperature rise of over 3°C by 2100. That might not sound like much, but it would be catastrophic, because this is an average that includes greater temperature rises where it matters. It would be impossible to live in large parts of the world. Many millions of people would die, if life survives at all. Crop yields would fall, and those who survive would fight over the remaining land and food, leading to many more deaths. That is the forecast for less than 80 years from now. Children alive today are facing a traumatic future if we do not turn this around.

Climate change attacks the building blocks we need to survive – air, water, food, health and security. Diseases, including pandemics, increase. We are all at risk. This is a crisis of choices that we have made and keep on making: greed, selfishness, waste, destruction. The climate is changing, plants and animals are dying, and we are heading for a dangerous future. As David Attenborough said, 'If we don't take action, the collapse of our civilisations and the extinction of much of the natural world is on the horizon.'[1]

Think of a child you know. Think about how old they will be in 2100, and any children they might have had. Think about the world they might be living in then. Will you do what you can to change the climate and stop this disaster?

Jamie will. On New Year's Day 2020, he set himself a New Year's resolution: 'To do more about climate change.' We wondered what that might look like.

His parents, David and Debbie, also made resolutions: not to fly, and to study the Bible more. (And to tidy the house.)

As a family, we started to study Bible passages that guided us in how to respond to the climate crisis. Jamie was very engaged with these, adding his opinions and ideas.

In 2019, Debbie had published a book with Tony Horsfall called *Resilience in Life and Faith.* Towards the end of the book, she discussed having a resilient environment. In early 2020, she felt prompted to ask David and Jamie whether they could, as a family, write a workbook about how to create a more **sustainable** environment. That same day, Olivia Warburton, commissioning editor at BRF, asked whether Debbie would consider writing a workbook as a follow-up to the resilience book. *Changing the Climate* was conceived.

A generation of young people are passionate about the climate. We want to help them to be passionate about the Bible too, by showing what the Bible says about creation care. We believe that everyone needs to know about climate change and that it's useful to know what the Bible has to say about it. *Changing the Climate* refers to every book in the Bible. Can you spot them all?

We want to challenge readers of all ages to act on the issue of **climate justice**. We have provided case studies and information about mission organisations, to encourage prayer and support for projects worldwide.

While this book was being written, much of the world went into lockdown because of the Covid-19 pandemic. We continued our ministry remotely, started online schooling and wrote chapters in free moments.

One of Jamie's early phrases when he was a toddler was 'global mourning', after he misheard the words 'global warming'. We need to do more than mourn. We must act while there is still time. Therefore, at the end of each chapter, Jamie provides tips about actions we can take. He has ranked each set of challenges from easiest (1) to hardest (12), in his opinion. You might give them a different ranking. We hope that you will take up many of the challenges.

David and Debbie are psychologists trained in understanding science. David spent a holiday studying the lengthy Intergovernmental Panel on Climate Change (**IPCC**) report on global warming of 1.5°C. He has contributed much of the scientific detail in this book. Debbie has taken the lead in providing Bible studies, based on discussions we have had as a family.

How to use this book

As well as welcoming individual readers, we also invite families, youth groups and house groups to discuss chapters together (see Appendix E). Each chapter can be read over several sittings. Teachers and home educators are welcome to use this material when teaching science, geography, religious education, life skills and other lessons.

We offer 'For deep thinkers' sections for those who like additional details. Skip these if you prefer to keep to the basics.

If you are not convinced that we should care about creation, start with chapter 3. If you want to understand climate change better, read Appendix A. If you don't understand the words in bold, look up Jamie's glossary at the back.

We are aware that many of the issues we discuss are more complicated than we have space to explore, and there are different opinions. We have listed our sources in the endnotes, so that you can check where the information came from and decide whether you trust the writer. You might not always agree with us, and we welcome disagreement. As we wrote this book, significant new research seemed to hit the headlines every week. Some details in this book may need to be updated in due course, but we don't think the main message will change. When the potential risk is huge, we need to act on the best evidence, not wait for complete certainty.

Our prayer is that as you read, you will put the actions into practice, so that *Changing the Climate* will help us all do more about **climate justice**. However you choose to use this book, we pray that it will help us all to do more to tackle climate change. Then at least some New Year's resolutions will have been accomplished. Even if our house is messier than ever!

PART 1

The problem

1 Waste

Bible passage

LUKE 15:11–32

Jesus continued: 'There was a man who had two sons. The younger one said to his father, "Father, give me my share of the estate." So he divided his property between them.

'Not long after that, the younger son got together all he had, set off for a distant country and there squandered his wealth in wild living. After he had spent everything, there was a severe famine in that whole country, and he began to be in need. So he went and hired himself out to a citizen of that country, who sent him to his fields to feed pigs. He longed to fill his stomach with the pods that the pigs were eating, but no one gave him anything.

'When he came to his senses, he said, "How many of my father's hired servants have food to spare, and here I am starving to death! I will set out and go back to my father and say to him: Father, I have sinned against heaven and against you. I am no longer worthy to be called your son; make me like one of your hired servants." So he got up and went to his father.

'But while he was still a long way off, his father saw him and was filled with compassion for him; he ran to his son, threw his arms round him and kissed him.

'The son said to him, "Father, I have sinned against heaven and against you. I am no longer worthy to be called your son."

'But the father said to his servants, "Quick! Bring the best robe and put it on him. Put a ring on his finger and sandals on his feet. Bring the fattened calf and kill it. Let's have a feast and celebrate. For this son of mine was dead and is alive again; he was lost and is found." So they began to celebrate.

'Meanwhile, the elder son was in the field. When he came near the house, he heard music and dancing. So he called one of the servants and asked him what was going on. "Your brother has come," he replied, "and your father has killed the fattened calf because he has him back safe and sound."

'The elder brother became angry and refused to go in. So his father went out and pleaded with him. But he answered his father, "Look! All these years I've been slaving for you and never disobeyed your orders. Yet you never gave me even a young goat so I could celebrate with my friends. But when this son of yours who has squandered your property with prostitutes comes home, you kill the fattened calf for him!"

'"My son," the father said, "you are always with me, and everything I have is yours. But we had to celebrate and be glad, because this brother of yours was dead and is alive again; he was lost and is found."'

The story of the wasteful son

This well-known story is often called 'the parable of the prodigal son'. A parable is a simple story with a spiritual meaning. 'Prodigal' means 'wasteful' or, more fully, 'spending money or using resources freely and recklessly; wastefully extravagant'.[1]

In this parable, the younger son asked for his share of his father's inheritance while his father was still alive, showing disrespect for his father. This was like saying to him, 'I wish you were dead so that I could have my inheritance.' Instead of continuing to live and work with his father, he cashed in his share of the money and travelled to a far country, where he 'squandered his wealth in wild living' (v. 13). To 'squander' means to waste in a reckless and foolish manner. The son spent everything he had, the whole inheritance, which should have lasted years. He had nothing at all to show for it.

The picture here is one of utter wastefulness and failure to think about the longer-term future.

Can you think of any ways in which you are wasteful?

Is there any way in which people in the world today are being wasteful in how they use up resources? If so, which resources?

After the son ran out of money, there was a severe famine. He got a job feeding pigs. A Jewish person who touched pigs was considered ritually unclean. Other Jews would not want to go near him. The son would have considered this a disgusting job, standing among pig droppings. An equivalent job today might be a sewage diver, swimming through sewage to clear out blockages.

The son came to his senses and went back to his father. He realised that he had sinned. The sin he is known for is being prodigal, or wasteful.

The younger son expected to return home as a servant. Instead, his father welcomed him as his beloved child and threw a party for him.

There are times when we, like the prodigal son, need to realise that our wastefulness is foolish, and even sinful. We, too, need to repent (which means to turn around and go in the opposite direction). We might need to say sorry, like the son did, and to make amends.

A wonderful part of this story is the forgiveness of the father. He was full of compassion for his son. He was so excited to see his son coming home that he ran and welcomed him, not caring that he looked undignified.

This is a story of hope, indicating that there is still time for us to change and to live differently.

The father represents God, who is merciful and forgives us when we turn away from wasting our lives and doing wrong. We need to do our part and not continue being wasteful.

It is easy to be like the older brother, who complains that it's unfair when his little brother is welcomed back. Little brother has already used up his inheritance, and now he is going to live off the inheritance that was meant for the older brother! In the next chapter we will consider contentment. The older brother was not content.

You might complain, 'I have spent years being careful and not wasting. I have spent years caring for creation. Now someone else is turning up late and wants to join the party and get the benefits.' Let's welcome everyone who makes positive changes, whatever they did before. We all need to work together, because there is a big job to be done.

Let us consider some ways in which we are wasteful.

Food waste

Humans worldwide waste about a third of the food we produce every year.[2] In 2015 in the UK alone, households wasted about five million tonnes of food – enough to fill 70,000 three-bedroom houses. That food was worth around £15 billion – almost £70 per month for an average family with children. Manufacturers, shops and caterers wasted another two million tonnes of food.[3]

Initial steps in reducing food waste include: not buying more than we need; putting only as much on plates as we will eat; and saving any leftovers to use later (or offering them to someone else), instead of throwing them away. See chapter 9 for more tips.

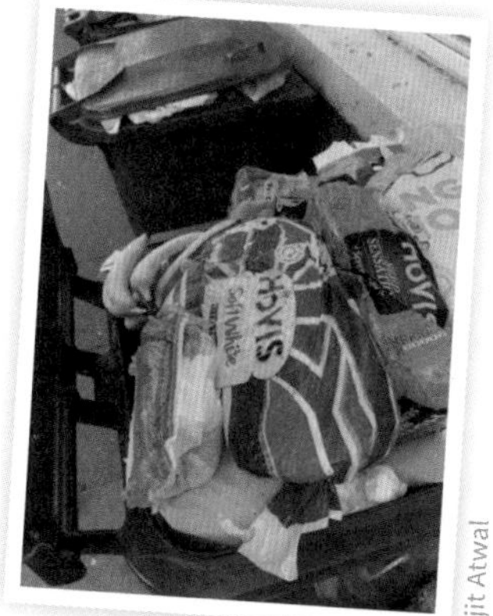

Ajit Atwal

Clothing waste

Fashion uses more energy than all planes and ships combined, and contributes to around 10% of global **greenhouse gas emissions.**[4]

People actively use the same clothes for about three years on average.[5] If we used clothes for longer, fewer would need to be made. That would be far better for the **environment**. Instead of buying new clothes to keep up with fashion, we can create our own fashion or just wear what we like and what we feel comfortable in (including second-hand clothes). We can wear the same items over and over again, buy clothes that will last and look after them.[6] If we want a new look, we can mix and match different clothes or alter garments, for example by tie-dying. This is not just for girls. Boys can also enjoy using environmentally friendly fabric crayons to decorate old white T-shirts.

If we grow out of clothing, we can pass it on to someone else, donate it to a charity shop or sell it or offer it for free online,[7] so that it can be reused. Items of clothing with holes in them can be mended or recycled as rags. Sadly, £140 million worth of clothing goes into landfill sites each year in the UK,[8] which is bad news for the **environment**.

We try to follow five Rs before throwing anything away (see Figure 1):

- ***Reduce*** the number of items we buy
- ***Reuse*** what we have (or let others reuse it)
- ***Repair*** items so they can be used again
- ***Recycle*** items when they are finished with
- ***Recover*** the energy from waste. For example, composting food waste and garden waste (grass and plant cuttings, and leaves) to recover the carbon. The compost can be put back into soil, improving its quality. Some non-recyclable waste can be used as fuel.

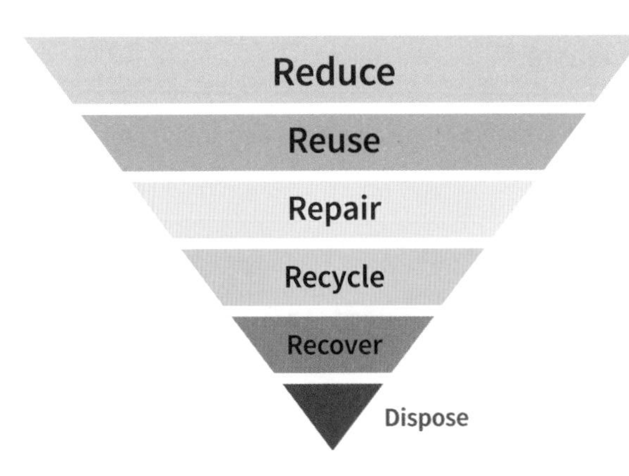

Figure 1 Five Rs

We should start at the top of this pyramid, with Reduce, and move down to the next step after doing all we can. Dispose only after trying the others first. This is not just for clothes, but for everything else as well.

What Jamie did

I wrote to the local newspaper to ask why we get free rubbish bins in our area but have to pay for recycling bins. I asked if it could be the other way round to encourage more recycling. A former councillor gave us a free recycling bin in response!

Plastic waste

Plastic is not all bad.[9] But we waste a lot of it, at a huge cost to the **environment**. Plastic production is expanding worldwide.[10] It contributes to **greenhouse gas emissions** (thus accelerating climate change) at every stage of its life cycle, from its production to its refining, and the way it is managed as a waste product.

Too much plastic is used once and then thrown away. Nearly all plastic can be recycled, but often it is not.[11] A lot of plastic collected for recycling has been dumped in countries like Malaysia, where it leads to pollution, causing breathing problems and other health issues.[12] According to one report, the plastics industry has deliberately exaggerated the possibility of recycling in order to sell more plastic![13] Recycling can work, but buying less plastic is far better.[14]

Hermes Rivera on Unsplash

Plastic waste in Nicaragua

Did you know?

The equivalent of one rubbish lorry of plastic enters our seas every minute.[15]

Eight million tonnes of plastics enter our oceans every year, adding to an estimated 150 million tonnes already in our seas. Plastic has been found in more than 60% of all seabirds tested,[16] and in about 50% of sea turtles.[17] Animals mistake plastic for food and eat it, and they may die as a result. Moreover, one person dies every 30 seconds from diseases caused by plastic pollution and other waste.[18] By wasting plastic, we are contributing to climate change and poisoning people and animals.

Did you know?

Most tea bags have plastic in them! If you like drinking tea, would you be willing to try loose-leaf tea (from a teapot, with a strainer) instead? Or try a brand of tea bag that does not use plastic. For example, Clipper say their tea bags are both Fairtrade and free from plastic.[19]

Wasting electricity and gas

Heating buildings contributes to at least 10% of the UK's **carbon footprint**,[20] and possibly much more. In 1970 the UK's average indoor temperature was about 12°C. Since 1997 it has been about 18°C.[21] To reduce **carbon emissions**, we could wear extra layers in cold weather and turn down the heating slightly.

Having a well-insulated building and efficient central heating is also important.[22] Turning off lights and electrical devices when they are not in use helps reduce wastage, although we should not fool ourselves that doing this is enough to solve the problem and fail to make more significant changes as well.[23]

Slavery: a waste of potential and opportunities

In the parable of the wasteful son, the younger son hired himself out to a pig farmer. He was so hungry he wanted to eat the pigs' leftovers, but was not allowed to. Though clearly he wasn't treated well, at least he had the freedom to go home, unlike a slave. Thankfully slavery is long gone. Or is it?

Modern slavery includes:

- children forced to pick food or cotton when they should be in school;
- forced labour;
- cocoa bean pickers who are exploited and controlled by someone else, without being able to leave;
- garment factory workers threatened with violence if they don't complete an order in time;
- car washers who have their passports taken away.

In the photograph below, this family in the Philippines were paid the equivalent of about 30 British pence for each large bunch of bananas they had collected from the mountains. Even the children performed a long, hard day of physical work (instead of attending school) in order to raise enough money to buy some rice to eat. This is child labour for a low wage.

Zeyn Afuang on Unsplash

Do you think any slaves work for you?

____

____

The Global Slavery Index[24] estimates that 40 million people are living in modern slavery, many of whom work in the supply chains of western clothing brands. These might be the people who make our clothes and shoes, pick ingredients for our food and drinks or mine for metal to make our mobile phones. Many of these workers are children, who are denied an education as they are forced to work, often in appalling conditions. We may be contributing to their wasted potential, as they are not paid appropriately or given adequate working conditions. This is a waste of human lives.

If you have access to the internet, try the survey at **slaveryfootprint.org** to estimate how many slaves work for you. You might be shocked by the answer.

Estimated number of slaves working for me:

___

Buying goods with a Fairtrade label[25] can help reduce the number of slaves working for you. To receive the Fairtrade label, companies need to pay an appropriate wage to workers. Fairtrade also sets social and environmental standards to protect workers, helping to promote better working conditions.

What else does the Bible say about waste?

When the Israelites wandered in the desert for 40 years, God provided them with a bread-like food they called 'manna' (which means 'what is it?'). God told them to collect an omer (thought to be about 1.4 kg) of manna for each person in their tent. Although some gathered too much and some too little, when it was measured, 'the one who gathered much did not have too much, and the one who gathered little did not have too little. Everyone had gathered just as much as they needed' (Exodus 16:18). Moses told them not to save any until morning. The first day some people disobeyed and saved some, and it turned smelly and became full of maggots. The people learned to gather just what they needed, and to eat it and not waste any.

This is a picture of a society where everyone had enough but not too much. After they got used to the system, no food went to waste. When Jesus taught his disciples to pray 'Give us today our daily bread' (Matthew 6:11), perhaps they remembered manna and that we only need enough for each day, not more.

Jesus fed about 5,000 men, plus women and children, with just five loaves of bread and two fish (Matthew 14:15–21). After everyone had eaten and felt satisfied, 'the disciples picked up twelve basketfuls of broken pieces that were left over' (v. 20). Later in a different place, Jesus fed 4,000 men, plus women and children, from seven loaves and a few fish. After everyone was satisfied, 'the disciples picked up seven basketfuls of broken pieces that were left over' (Matthew 15:37). Why do you think they bothered to pick up the broken leftovers on both occasions? After all, everyone had eaten enough, and they had started out with almost nothing. It must have taken quite some time to collect broken pieces left by thousands of people, so why would they bother? Does it suggest that it is good not to waste resources?

Is there a place for generosity?

How can we get a balance, so that we do less damage from wastefulness but are not mean or miserly?

In feeding the crowds so well and then having plenty of leftovers for people to eat later, Jesus showed how generous God is. Another time, when a woman poured a bottle of perfume over Jesus, he did not agree with critics who called that wasteful, but he commended her for beautiful generosity (Mark 14:3–9).

Returning to the parable of the wasteful son, the father seemed to have a good balance. His general pattern was not one of waste. He had saved money for his sons' future. The older son points out that he did not generally kill animals for parties. The calf had time to get fat. However, the father was generous. He gave his hired men plenty to eat (Luke 15:17). He provided a robe, ring and sandals for his returned son. He killed the calf to provide for a feast (as celebrating with meat was their tradition).[26] We guess that none of the meat would have been wasted, and probably the skin would have been used for clothing as well. The hired workers were invited to the celebration, with the father saying to them, 'Let's have a feast and celebrate' (v. 23). When the older brother asked them what the music was for, they knew because they were at the party.

We too should seek a balance of showing generosity in sharing with others, while not wasting resources.

Is there anything you want to ask God to forgive you for wasting?

____

____

____

Generous God,
in a world where many people are hungry or homeless, we are sorry that we waste food and resources. In a world where people are suffering and dying because of pollution and climate change, please forgive us for contributing to the problem. Like the prodigal son, we want to change, be forgiven and make better choices. Instead of wasting time, help us make the most of every opportunity (Ephesians 5:16). Amen

Loving our neighbours worldwide: Nuevas Esperanzas in Nicaragua

nuevasesperanzas.org

The prodigal son experienced a famine. From the earliest times, there have been times of famine, when there has not been enough food (Genesis 12:10). Today, climate change is increasing the chances of harvest failure on a devastating level.

As well as a shortage of food, a shortage of water is common in many parts of the world. Nicaragua has been badly affected by climate change.[27] Long dry seasons have led to crop failures. Many people lack the clean drinking water that we take for granted. In wet seasons, flash flooding causes further problems. The Global Climate Risk Index indicates that between 1997 and 2016, only three countries in the world had greater losses (including fatalities and economic costs) due to weather-related events than Nicaragua.[28]

Nuevas Esperanzas

Rainwater harvesting tank and tap

Nuevas Esperanzas (meaning 'new hopes') is an ecumenical charity working in Nicaragua. One of their projects involves helping local people to build huge ferro-cement tanks to capture and store rainwater which can then be used by communities who previously travelled for hours in search of water. With the rainwater harvesting tanks, not even rainwater is wasted. The charity also runs other projects: training community members in beekeeping; helping them to grow a variety of vegetables to diversify crops and income; building latrines; and supporting reforestation.

> **'God must think we are crazy. He sends the water through rain and we let it run off down the hill, and then we go far downhill to look for it.'**

Young people around the world: Fetching water in Nicaragua

Nuevas Esperanzas

Adiel Joel Rivera, 8, has lived all his life in Agua Fría, the hillside community closest to the crater of the active Telica Volcano. Adiel enjoys school and is learning to read. He loves living in this beautiful location. It is cool, and he loves to play. This sounds idyllic, but life on the side of a volcano has its challenges as well. Adiel's father described how hard it can be to get down the mountain to find medical care if one of his children is sick or to evacuate when the volcano erupts. When the rains come, it is even harder. When there is too much or too little rain, or acid rain (created by gases from the volcano), crops can be damaged, causing the harvest to fail. If this happens, Adiel's father is forced to leave home for several months to find a job in the neighbouring country of Costa Rica.

Until Nuevas Esperanzas built a rainwater harvesting tank close to Adiel's home, the family had to spend two hours every day making trips to fetch water from the nearest spring. This meant having a horse and borrowing containers to fill with sufficient water for drinking, washing, bathing and cooking. Adiel sometimes went to the spring to fetch water. His dad always made sure he went on horseback as it was so far from home. The rainwater harvesting tank has made a big difference, providing clean water near their home. This means there is one less thing to worry about in the challenges of life on the side of the volcano.

> **The poor and needy search for water, but there is none; their tongues are parched with thirst. But I the Lord will answer them; I, the God of Israel, will not forsake them.**
> ISAIAH 41:17

Jamie's tips

1. Avoid buying plastic bottles of water. Instead, use a reusable bottle. If you run out of water, you can ask for it to be refilled at motorway service stations, coffee shops or cafés.

2. Try to reduce the amount of plastic you acquire. Take your own bag to shops, so you don't need to accept a plastic bag (even if it's free). Try to buy items without plastic packaging. You could even bring your own container when you get takeaways.

What Jamie did

I created a PowerPoint presentation about plastic waste and showed it at my youth group. We discussed how to reduce our use of plastic.

3. Try to choose some food that would otherwise be thrown away. For example, some shops sell 'wonky' vegetables or chips (which have unusual shapes) or broken biscuits. Many supermarkets sell items at a reduced price when they are close to the end of their shelf life.[29] Unsold food is often thrown away. Sign up to Olio (**olioex.com**), an app for sharing unwanted food and other items.

4 If you can, hang clothes to dry outside instead of using a tumble dryer, so you don't waste electricity. Running a clothes dryer uses as much electricity as 225 low-energy light bulbs.[30]

David Hawker

5 When you finish with clothes, books, games or other items, offer them to a friend or take them to a charity shop.

6 Find ways to recycle items that are usually thrown away. If you cannot make them into something new yourself, look for companies which can use them.

Did you know?

Terracycle pays schools, organisations and individuals for collecting used items, such as crisp packets, pens, toothbrushes, toothpaste tubes and contact lenses. Terracycle makes the waste into tables, benches, play equipment, bike racks and other products. Terracycle operates in the USA, UK, Republic of Ireland and ten other European countries.[31]

7 Spend less money on yourself this month. Donate what you save to rainwater harvesting and other projects in Nicaragua. See **nuevasesperanzas.org**.

8 When you buy a phone, choose either a second-hand one or a Fairphone[32] (which uses recycled and Fairtrade materials and is designed to last). Encourage your family and friends to send old phones to the Fairphone company to be recycled.

9 Collect clean, dry plastic which cannot be recycled and push it into an empty plastic bottle. This can be made into an ecobrick, which can be used for building furniture, play parks and other structures. See **ecobricks.org**.

Timothy Lonsdale, 7, makes ecobricks from plastic waste, assisted by sister Alice, 4

10 Consider switching to a renewable electricity supplier[33] or encourage your church to do so. Also consider how energy-efficient your home/church is and whether this could be improved, such as with better insulation.[34]

11 List everything you waste, including fuel, electricity, water, food, clothing, paper, plastic and other items. Try to reduce your waste as much as you can.

12 Watch Tom Bray's YouTube video on **sustainable** fashion[35] and try not to buy any new clothes for yourself for six months.

Which of these do you already do?

____

____

____

Which will you attempt to do this month?

____

____

____

2 Want

Bible passages

EXODUS 20:17

'You shall not covet your neighbour's house. You shall not covet your neighbour's wife, or his male or female servant, his ox or donkey, or anything that belongs to your neighbour.'

PHILIPPIANS 4:11–13

I am not saying this because I am in need, for I have learned to be content whatever the circumstances. I know what it is to be in need, and I know what it is to have plenty. I have learned the secret of being content in any and every situation, whether well fed or hungry, whether living in plenty or in want. I can do all this through him who gives me strength.

Waste not, want not

The phrase 'Waste not, want not' means that if you use what you have carefully, you will still have some if you need it in the future. In other words, if you don't waste, you won't be found wanting later.

'Want not' could also mean 'don't want what you don't have – be content with what you have and don't continually desire more things'.

Do not covet

'You shall not covet' is one of the ten commandments God gave the Israelites (Exodus 20:17). It means don't desire things that don't belong to you. Wanting can be a problem.

They are not ten suggestions. They are commands from God, like ‘Do not murder’ and ‘Do not steal’. Jesus said we should keep them; they are not just for the Old Testament (Matthew 19:17–19). The apostle Paul wrote that ‘You shall not covet’ is one of the commandments summed up in the rule Jesus taught, ‘Love your neighbour as yourself’ (see Romans 13:9).

Exodus 20:17 lists specific examples of things that we should not covet: ‘You shall not covet your neighbour’s house... [or] wife, or... servant, his ox or donkey.’ Perhaps by this point in the list, you think you are doing well. But then the verse finishes, ‘or anything that belongs to your neighbour’.

Maybe you don’t envy anyone’s ox or donkey. But do you ever envy their bike or car? Do you ever envy someone’s clothes, phone, smartwatch, other belongings, wealth or holiday? Do you wish you had a better house, newer technology or better parties or days out?

Luke Chesser on Unsplash

What do you desire that does not belong to you?

____

____

____

____

Coveting leads us to try to use up more resources than we have. During the Covid-19 pandemic, after initial panic-buying, economies ground to a halt as people bought only what they needed. What kind of an economy do we have that relies on us buying what we don’t need?

God’s people were instructed not to keep all the harvest for themselves, but to leave some for the poor (Leviticus 19:9–10; see also Obadiah 5; Ruth 2:15–16).

A rich young ruler who had kept all the commandments asked Jesus what else he must do. Jesus told him to sell his possessions and give the money to the poor (Luke 18:18–23). Selling all our possessions is not wasteful, as throwing them away would be, but usually means settling for a price much lower than we bought them for. It is a reversal of covetousness.

Here are challenges for us:

- Share our resources with people who need them.
- Stop getting so many new things.
- Give away things we don't need.
- Consider how our greed affects people far away and in the future.

Did you know?

Strangely, sacrificing one luxury can make us think we are justified in consuming another. For example, people who eat less meat because of climate change might use the money saved to fly more. This is called the Jevons paradox, and often stops us making any progress regarding climate change. Can we avoid making that kind of mistake?

Be content

Instead of coveting, the Bible teaches us to be content. In our second passage, the apostle Paul writes, 'I have learned the secret of being content in any and every situation, whether well fed or hungry, whether living in plenty or in want' (Philippians 4:12).

What do you think this secret is?

____

____

____

____

Part of Paul's secret was that he relied on God to give him the strength he needed (v. 13). Also, Paul wrote, 'Set your minds on things above, not on earthly things' (Colossians 3:2). Likewise, Jesus taught:

> Do not store up for yourselves treasures on earth, where moths and vermin destroy, and where thieves break in and steal. But store up for yourselves treasures in heaven… For where your treasure is, there your heart will be also.
>
> MATTHEW 6:19–21

Simplicity

Paul wrote that he had learned to be content even when in need. He (like Jesus and his disciples) was used to travelling light. He even seemed content in prison – he was in custody when he wrote these words (see Philippians 1:13–14; see also Acts 16:25; Philemon 1–7; 2 Timothy 1:8). Paul lived a simple lifestyle. He encouraged the Philippians not to complain (2:14) but to be thankful (4:6).

Did you know?

Making time to be thankful tends to make people more happy, joyful and optimistic, and to help them sleep better and feel more connected with other people and closer to God.[1]

Embracing simplicity, and being thankful for what we have instead of complaining about what we do not have, may be part of the secret of contentment. People who don't have many possessions often find joy in small things. While many children in Britain complain about not having the latest technology, we have watched children in Africa play joyfully with broken toys, household objects, items they have made from rubbish, or sticks and stones. We have also seen children express delight in having materials to draw a picture, while not complaining about wearing torn clothes.

David Hawker

Debbie with children at a school in Ethiopia

Debbie met a woman in El Salvador whose small one-room hut was regularly flooded. She and her children had almost no material goods, but she said with a smile, 'We are rich because we have God.'

Making comparisons or being grateful?

If we make comparisons, why not compare ourselves with people who have less than we do, not more?

What Jamie did

When Jamie was two, we used to say 'Thank you, God' prayers together every evening. He would thank God for all sorts of things which we had never thought of thanking God for – for example, each of the London underground lines and brushing teeth. Taking time to be thankful helps us to feel content instead of always wanting more. When, as a toddler, Jamie received a Christmas stocking, he took a long time to examine every item in it, showing delight at every little thing, like a coloured pencil and a packet of crisps. He taught us to be more grateful for what we have, instead of wanting more.

> Two things I ask of you, Lord; do not refuse me before I die: keep falsehood and lies far from me; give me neither poverty nor riches, but give me only my daily bread. Otherwise, I may have too much and disown you and say, 'Who is the Lord?' Or I may become poor and steal, and so dishonour the name of my God.
>
> PROVERBS 30:7–9

If you live in a developed country, you're richer than most of the world. With a lot of wealth, we can start to rely on ourselves and what we have, and forget about God. Missionaries have told us that they have been more content and relied more on God when they had little money.

What does this have to do with climate change? Figure 2 (on the next page) is a rough illustration of how our greed drives the environmental crisis and damages the climate.

Throughout this book we will refer to the middle and bottom sections of Figure 2, and what we can do to reduce the harm. But in this chapter, we want to consider the top section, which is perhaps the biggest cause of climate change.

Do you agree with Figure 2? What do *you* think are the root causes of the environmental crisis?

____

____

____

We believe that a key cause of climate change is that people continually want more – more food, clothes, plastic stuff, technology, heating, cars and travel than we really need. To provide what we want, we harm the world God has made – cutting down forests and increasing **greenhouse gas emissions**, plastic waste and pollution.

As a result, creation is in decay (as Paul wrote in Romans 8:21). The world gets warmer (see Appendix A); there are more **droughts**; ice melts, the sea levels rise and there is more flooding. This causes more **climate refugees**, the loss of wildlife and the spread of disease.

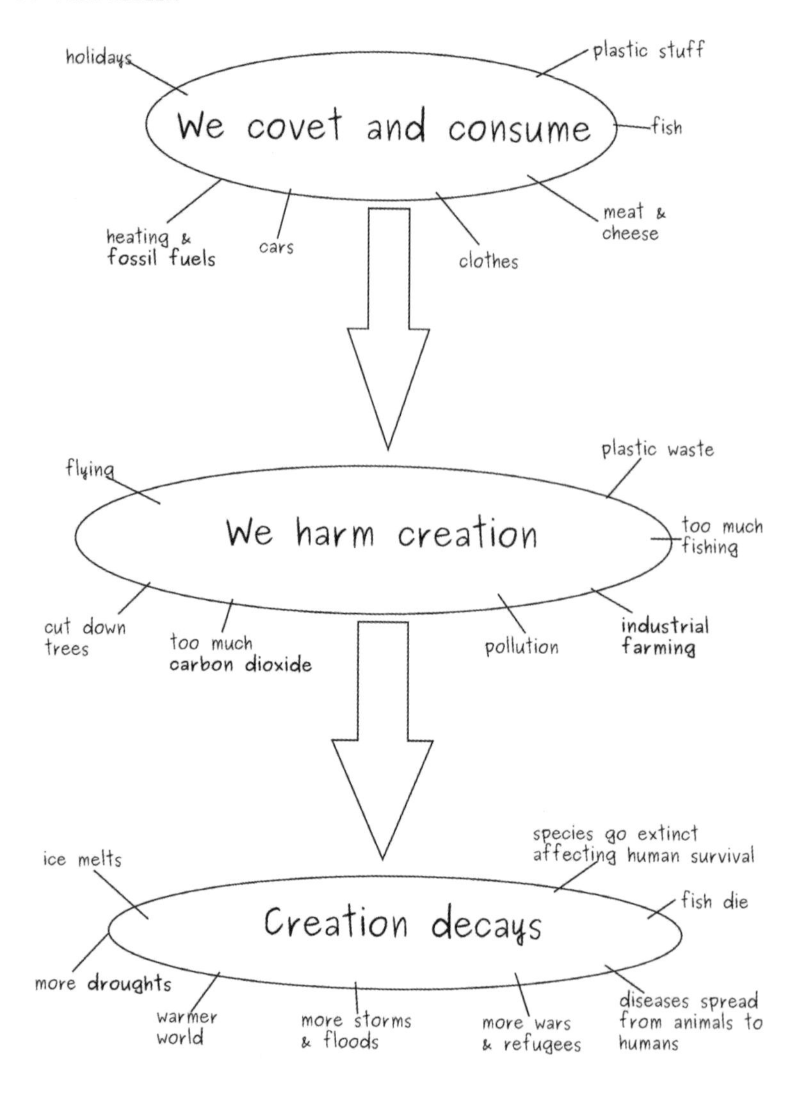

Figure 2 Creation in crisis

Creation waits for the children of God to be revealed (Romans 8:19) – the ones Jesus called peacemakers (Matthew 5:9), those who make everything all right. Will we reduce the decay of creation? Being content with a simpler lifestyle, and therefore reducing consumption, can help, as shown in Figure 3.

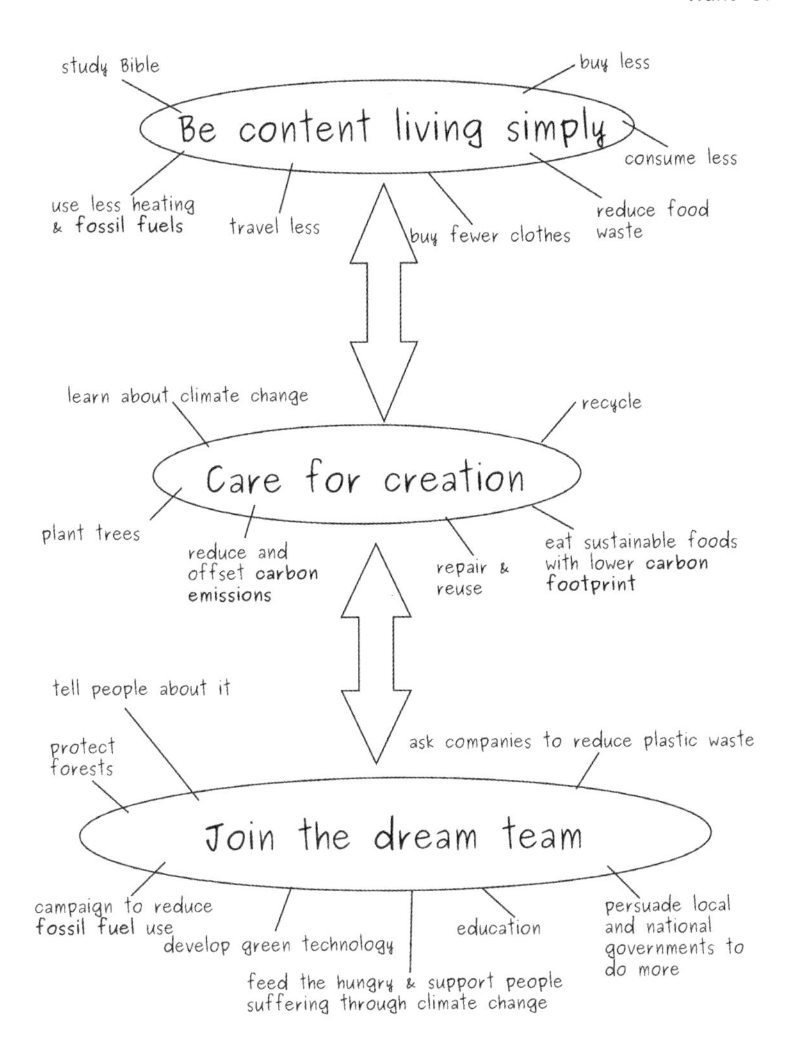

Figure 3 Creation awaits the children of God

To put things right, we can try to be content with what we have, reduce our consumption and encourage others to do the same. We can encourage companies to reduce the amount of **fossil fuels** and plastic they use and the waste they produce. We can also support those suffering the effects of climate change, for example by praying for, donating to or volunteering with the charities featured in this book.

What young people are doing

Sisters Ella and Caitlin McEwan, 9 and 7, started a petition to call on McDonalds and Burger King to stop putting plastic toys in children's meals, as the toys are often thrown away unused. Over 568,000 people signed the petition. Burger King agreed to stop the toys, while McDonalds agreed to give the option of a toy/book or fruit or carrots.

Prayer

Lord Jesus,
you stepped down from heaven to be born into poverty and became a refugee for our sake. Please help us to be content and not to store up treasure on earth. Help us to live simply and to care for your creation. May we become peacemakers as true children of God. Amen

Loving our neighbours worldwide: Christian Aid

caid.org.uk

Mahatma Gandhi is reported to have said, 'Live simply so that others may simply live.' For decades, Christian Aid has taken up this challenge, calling people to live simply.

Among other activities, Christian Aid hosts petitions demanding that governments pursue **climate justice**. Christian Aid partners help to provide renewable energy in countries such as Guatemala, the Philippines and Malawi.

Christian Aid

Climate Change Advocacy Group member, Kenya

In Kenya, Christian Aid provides guidance about climate change planning, as well as helping communities to access **sustainable** energy. Climate change is already adversely affecting Kenya through increased temperatures, **drought** and erratic rainfall, so doing something about it is very important.

Young people around the world: From fossil fuels to clean energy in Kenya

Christian Aid

Since starting secondary school, Janet Machuka has been a 'young activist' with Christian Aid, involved with climate change projects. She explains:

> I was born in a village where we could only afford to go to the forests and fetch firewood for fuel. We never knew what electricity was. Paraffin was our source of light as we lit our grass-thatched houses with the little lamps. The sun was only used for drying our maize. We never knew it could be used to produce electricity, so we continued our habits of cutting down trees to light and warm our houses.
>
> Dark smoke from the fuels left us constantly visiting the hospital as my mama is asthmatic, and the smoke from the firewood and coal impacted her health every time she cooked.

As well as these health problems, these forms of energy use were bad for the **environment**. Cutting down trees stops them from absorbing carbon, burning wood produces **carbon dioxide** and paraffin is a **fossil fuel**. Janet continues:

> In primary school we learnt about renewable and non-renewable sources of energy. Back then, those lessons were just for the classroom. We continued to live our day-to-day lives while our health continued to be impacted by the **fossil fuels** in our village. But as we grew older, we took those lessons more seriously and made a conscious decision to invest in the necessary changes to our everyday lives. Now, my mama uses energy from a solar panel to light our homes at night, and prepares our meals in a cleaner **environment**.

Jamie's tips

1. Listen to (or sing) the song 'God of the poor (Beauty for brokenness)' by Graham Kendrick. You can find this on YouTube. Pray the words of verse four.

2. Write out and try to memorise 1 Timothy 6:6–8: 'But godliness with contentment is great gain. For we brought nothing into the world, and we can take nothing out of it. But if we have food and clothing, we will be content with that.'

3. Set up a 'gratitude pot' for your household. Every time you think of something to be thankful for, write it down and put the note in the pot. At the end of the month, read through all the notes and take time to thank God.

4. Think of as many Bible verses as you can about greed, riches or contentment. If you can't think of any, read Matthew 19:21–26; Luke 12:15; Hebrews 13:5; 1 John 2:15–17.

5. Instead of buying new items, try to get them second-hand. This reduces the amount of energy needed to produce new things. This goes for both expensive goods, such as computers and bikes, and cheaper items, such as clothes and games.

Hints from the Hawkers

We like to give and receive gifts from charity shops. You can get more presents for the same price, the money goes to a charity and the carbon footprint is reduced compared with new goods having to be produced. Everybody wins! Alternatively, sometimes we ask for a donation to a charity instead of getting more stuff.

6 Get involved with Christian Aid's climate change campaign. You can sign petitions, read information, join events or receive prayer requests. See caid.org.uk/climate. Download a free poster about how climate change affects children around the world at christianaid.org.uk/sites/default/files/2017-08/crazy-climate-poster.pdf.

7 Send a donation to Christian Aid. Perhaps your group could arrange a 'rich person, poor person' meal, in which people are given a number when they book. Everyone receives a simple meal (such as soup and bread), but one-quarter of the people are randomly chosen (through picking their numbers out of a hat) to receive a 'rich person's meal' (such as a delicious dessert as well). Warn them beforehand that this will happen. Everyone is asked to donate to Christian Aid for their meal. A 'rich person, poor person' meal illustrates the injustice in the world where some of us have so much, while many people have so little.

8 Try hard to be kind to people at home/school/university/work this week. Focusing on being kind to people might help you feel more content. It might also help *them* to feel more content, which means that they might not feel the need to have lots of extra things.

9 If you go clothes shopping, decide in advance not to buy lots of clothes. Choose just one or two new items that you really like and that you will be able to keep using again and again for a long time.

10 Try to go for a week without buying anything apart from necessities, such as meals. If a week seems easy, try a month.

11 Try not to complain this week. Instead, pray for people who aren't kind to you, and pray about the things that irritate or upset you. Choosing not to complain can help us become more content.

12 Talk to the people you go on holiday with about ideas for a holiday in the future that does not cost the earth – maybe staying nearer to home than usual or travelling by train or coach.

What Jamie did

Wild camping means camping in the wild without a campsite. I went with my dad and his friends and their children to Dartmoor in Devon, where wild camping is legal. We went there by train. Each day we walked to a new place to camp, enjoying the countryside and carrying our tents, which we put up each night. It was very good. I am hoping to do it again in Scotland.

Steve Gannon

Jamie (right) with friends, wild camping

Which of these do you already do?

____

____

____

Which will you attempt to do this month?

____

____

____

PART 2

Why we should act

3 Reasons to care for creation: faith in action

Should we care about the environment? Why, or why not?

If you are already convinced that we need to care for creation, feel free to skip quickly through this chapter. But maybe you've heard that if you're a Christian, it'll work out fine in the end. God has promised not to repeat the destruction of the great flood, so how could we destroy the world or human life on it? We have heard that this life is just a rehearsal, and our home is not on earth but in heaven. Why should we care about fallen creation?

If you're unsure, or want ideas to help persuade others, see what you think of the following eight reasons to express our faith by caring for creation.

Bible passages

GENESIS 1:24–28

And God said, 'Let the land produce living creatures according to their kinds: the livestock, the creatures that move along the ground, and the wild animals, each according to its kind.' And it was so. God made the wild animals according to their kinds, the livestock according to their kinds, and all the creatures that move along the ground according to their kinds. And God saw that it was good.

Then God said, 'Let us make mankind in our image, in our likeness, so that they may rule over the fish in the sea and the birds in the sky, over the livestock and all the wild animals, and over all the creatures that move along the ground.'

So God created mankind in his own image,
in the image of God he created them;
male and female he created them.

God blessed them and said to them, 'Be fruitful and increase in number; fill the earth and subdue it. Rule over the fish in the sea and the birds in the sky and over every living creature that moves on the ground.'

GENESIS 2:15, 19

The Lord God took the man and put him in the Garden of Eden to work it and take care of it...

Now the Lord God had formed out of the ground all the wild animals and all the birds in the sky. He brought them to the man to see what he would name them; and whatever the man called each living creature, that was its name.

Eight reasons to care for creation

1 Respect for the creator

Genesis 1 says that God created the earth, plants, animals and humans, and saw that they were good. God created a good creation. It is right to respect it.[1]

It is said that Leonardo da Vinci was once close to finishing a painting he had been working on for months, when he offered his paintbrush to a student and said, 'You finish it... Will not what I have done inspire you to do your best?'[2] How would da Vinci have felt if the student had deliberately ruined the masterpiece, pouring oil and food over it and throwing it away? The artist would rightly have felt saddened and aggrieved.

Now consider God the artist. God created an amazing world, with glorious sunsets, exquisite snowflakes, delicate flowers, awesome waterfalls and a huge variety of beautiful butterflies, birds and other animals. As the psalmist says, 'The earth is the Lord's, and everything in it' (Psalm 24:1).

Verma Langrell

Creation displays the nature and power of God (Psalm 19:1–2; Romans 1:20). Creation was made through and for Jesus (Colossians 1:16). God handed us the task of taking care of this amazing creation and adding our brushstrokes to the picture. We respect God by caring for creation.

What are some of your favourite parts of God's creation?

____

____

____

____

____

____

____

Did you know?

A third of all plant and animal species present in 2020 are likely to be extinct by 2070, owing to man-made climate change. Scientists say that we are in the sixth mass extinction, with vertebrate species disappearing at 100 times the natural rate.[3]

2 God cares about creation

Alison Walker

Seven times Genesis 1 says that God saw creation was 'good'.[4] God brought the animals to Adam to name each one; they were that important. Perhaps God was teaching Adam to value them individually, each worthy of a name, from the smallest to the largest.

Jesus told his followers that God cares for every little bird: 'Are not two sparrows sold for a penny? Yet not one of them will fall to the ground outside your Father's care. And even the very hairs of your head are all numbered' (Matthew 10:29–30).

Did you know?

In the UK, the number of tree sparrows has fallen by about 90% since 1970, and the number of house sparrows by over 50%.[5]

In Psalm 50, God declares:

> For every animal of the forest is mine,
> and the cattle on a thousand hills.
> I know every bird in the mountains,
> and the insects in the fields are mine.
>
> PSALM 50:10–11

God knows and cares about every bird, insect and animal. God cares deeply about creation. We are created in God's likeness. Following God means caring about creation.

3 Failing to care for creation has devastating consequences

> **Justin Welby, Archbishop of Canterbury, said in 2019:**
> **'It becomes ever clearer that climate change is the greatest challenge that we and future generations face. It's our sacred duty to protect the natural world we've so generously been given, as well as our neighbours around the world who will be first and worst affected. Without swift decisive action the consequences of climate change will be devastating.'[6]**

The United Nations Intergovernmental Panel on Climate Change (**IPCC**) includes hundreds of scientists from 195 countries. Repeatedly the **IPCC** has confirmed the climate is changing because of human activity. We wrote this chapter on the day of the death of Sir John Houghton, a leading climate scientist and a Christian. He was the co-chair of the **IPCC** and the former chief executive of the Met Office (the UK's national weather service). He wrote:

> The impacts of global warming are such that I have no hesitation in describing it as a 'weapon of mass destruction'. Like terrorism, this weapon knows no boundaries. It can strike anywhere, in any form – a heatwave in one place, a **drought** or a flood or a storm surge in another. Nor is this just a problem for the future... Global warming is already upon us.[7]

Dozens of countries have declared a climate emergency, but the amount of **carbon dioxide** in the atmosphere continues to rise. The need for action is urgent (see Appendix A).

Up to 2020, the five warmest years on record were 2015 to 2019. The tropics have experienced deadly heatwaves. Sub-Saharan Africa is seeing heavier rains with longer dry spells in between – a disastrous combination for harvests. Ice is melting in Antarctica, Greenland and mountain glaciers. Melting land ice raises the sea level. Melting sea ice stops reflecting sunlight and absorbs it, warming the sea. As the sea warms, it expands and the sea level rises further. Higher temperatures increase evaporation and consequently rainfall and flooding. The actions of humans today will have a huge impact on what sort of earth remains for the next generation.

Did you know?

The Republic of Maldives is a small island nation in South Asia. Former President Mohamed Nasheed tells us, 'The Maldives are disappearing. We will soon be under water... climate change is already upon us... erosion and water contamination are getting worse... The coral reefs that protect our islands and provide us with food are threatened with extinction by ever-higher ocean temperatures that bleach and kill coral. Then there is the longer-term threat of submersion by the rising seas – the highest point in the Maldives is just six feet above water.'[8]

An IPCC report confirmed that small islands, including the Maldives, the Marshall Islands, Kiribati, Tuvalu and many others, are likely to disappear due to climate change.[9]

Some see the climate emergency as a liberal hoax or a 'leftie agenda', pushing people away from trusting God towards trusting science or a political party. But climate change deniers are now in a small minority. A survey of 30,000 people in 28 countries found that only 0–6% of people in any nation deny that climate change is happening (including 2% in the UK and 6% in the USA).[10] A majority of people surveyed think climate change will cause the destruction of cities due to rising sea levels, seriously damage the overall economy, lead to large numbers of **climate refugees** and cause wars.

In biblical times, the climate was repeatedly linked to politics, sin and God's judgement (e.g. Genesis 6—8; Amos 4:6–8; Haggai 1:10–11). When we use the minds that God has gifted us, we can see the evidence that the climate is changing and that the consequences are likely to be devastating. Our response can draw us closer to God as we ask God to guide us in acting out of love for our neighbours.

Although the impact of climate change is generally worst in low-income nations, nowhere is immune. The year 2020 began with bush fires burning Australia, then Siberia and California burned in the summer. In February 2020, parts of Britain were flooded for the third time in a fortnight after unusually fierce storms. The same pattern is being repeated worldwide.

Which aspect of climate change are you most concerned about?

____

____

____

____

4 Love for neighbours and future generations

Chris Gallagher on Unsplash

Worldwide, climate change is causing an increase in flooding, **droughts**, food insecurity, health problems and many other crises. The World Health Organization (WHO) states that at least 250,000 people are likely to die every year between 2030 and 2050 as a result of health problems caused by climate change (malnutrition, malaria, diarrhoea and heat stress),[11] before adding in all the deaths due to floods and other natural disasters.

The apostle Paul urged us to offer ourselves as a living sacrifice (Romans 12:1). Making sacrificial choices to help limit climate change shows love for other people, including those who will come after us. The co-chair of the **IPCC** Working Group II has said, 'The decisions we make today are critical in ensuring a safe and **sustainable** world for everyone, both now and in the future.'[12]

The sixth commandment is 'You shall not murder' (Exodus 20:13). Our action or inaction related to climate change may cause death. If we continue to live in a way that accelerates climate change, without caring about those who will suffer and die as a consequence, then we are contributing to their pain and deaths. Jesus said, 'Love your neighbour as yourself' (Mark 12:31), and he made it clear that everyone is our neighbour (Luke 10:36). Do we think the lives of our children and grandchildren (or others) are worth less than ours? Caring for the **environment** in a way that helps to limit the devastating impact of climate change is one way of loving our neighbours.

5 Justice

The Old Testament mentions justice and related words over 900 times. It means looking after people who have little power. The **ecological** crisis is severely unjust. Our grandchildren will pay for the damage we have done. Even now, countries which have caused the least climate change are often the worst affected.

For instance, Burundi has the worst food security in the world despite very low **carbon emissions** (see Appendix B). On average, in one year:

- one person in the UK produces the same emissions as 212 people in Burundi;
- one person in the USA produces the same emissions as 581 people in Burundi;
- one person in Australia produces the same emissions as 609 people in Burundi.[13]

Rich countries have generally caused more climate change but will suffer less from it. They also have the money to help them do more about it. Nick Spencer and Robert White write:

> People in high-income countries who have benefited historically from the burning of **fossil fuels** have a particular responsibility to deal with the results of their actions. And they can hardly claim to love their (global) neighbour when the consequences of their lifestyles may lead to suffering and an increased probability of an early death by people elsewhere. [That is] nothing less than sinfulness... putting oneself first... We need not only to repent of that but also to change our behaviour.[14]

6 God commands us to care for the earth

> **Archbishop Desmond Tutu said:**
> **'Who can stop climate change? We can. You and you and you, and me. And it is not just that we can stop it, we have a responsibility to do so that began in the genesis of humanity, when God commanded the earliest human inhabitants of the Garden of Eden, "to till it and keep it". To "keep" it; not to abuse it, not to make as much money as possible from it, not to destroy it.'[15]**

God's first command to humanity, in Genesis 1:28, was 'Be fruitful and increase in number; fill the earth and subdue it' and rule over the creatures. Does that give us permission to plunder and destroy the earth's resources? No. We are commanded to subdue (*kabash* in Hebrew) the earth because it is wild, and we need to tame or manage it if we're going to survive.[16]

The word translated as 'rule' over the earth (*radah*) is the same word that describes the just and peaceful rule in the early days of King Solomon (1 Kings 4:21–24). It 'reflects the character and values of God's loving and compassionate kingship'.[17] In *The Message*, the command is written as: 'Be responsible for fish in the sea and birds in the air, for every living thing that moves on the face of the Earth.' This idea of responsibility, or stewardship, fits well with the rest of the biblical narrative. The land belongs to God, and we are the tenants looking after it (Leviticus 25:23).

The land and its produce are intimately connected to our survival and bound up with worship. Throughout the Bible, the creation praises its creator. God's people rejoice in the harvest and give some of it back to God. Creation is central to worship. God put Adam in Eden to 'work it and take care of it' (Genesis 2:15). The Hebrew words here, *abad* and *shamar*, are used throughout the Bible for serving God and obeying God's commands. Our duty of serving and preserving creation is not an optional extra. It should be central to our worship.

Practices like a weekly sabbath, tithing and jubilee years showed the people of Israel that they were not to exhaust the earth or extract every available resource from it.

How well do you obey the command to care for God's creation? What more could you try?

____

____

____

7 An act of witness

Isn't bringing people to Christ more important? Don't we need to save people from eternal damnation rather than saving the planet and its inhabitants from destruction?

Holly-Anna Petersen/ Christian Climate Action

'Creation cries out' – banner held by former archbishop Rowan Williams and other clergy

Jesus commanded us to make *disciples* of all nations (Matthew 28:19), not converts only. A disciple follows Jesus and learns from him. 'Let your light shine before others, that they may see your good deeds and glorify your Father in heaven' (Matthew 5:16). Protecting the **environment** is an act of witness, demonstrating our faith through action. We can seek opportunities to tell people why we care about creation, as we share our faith.

8 Working in unity

Psalm 133 speaks of the blessing which comes from being in unity. Many Christian denominations have taken a stand on caring for the **environment**.[18] The box below includes a few examples, but there are more. Christians from different churches and denominations are working together for **climate justice**. Working together in unity is an act of witness.

Did you know?

- In 2006 nearly 100 key evangelical leaders signed a document entitled 'Climate change: An evangelical call to action'.
- In 2008 Southern Baptists issued a 'Declaration on the environment and climate change'.
- In 2009 the Vineyard churches agreed a series of measures to promote creation care.
- In 2019 the Pope declared a climate emergency.
- The Salvation Army is part of The Climate Coalition, calling for climate action.
- The Methodist Conference has recognised a climate emergency and called on the UK to achieve **net zero emissions** well before 2050.
- In 2020 the Church of England General Synod passed a motion calling 'upon all parts of the Church of England... to work to achieve year-on-year reductions in emissions and urgently examine what would be required to reach **net zero emissions** by 2030'.[19]

What has your church/denomination said and done about climate change?

____

____

____

The creation covenant

Bible scholar Margaret Barker, among others, argues there are more fundamental reasons in the Bible for caring for creation. Drawing not just on the Bible but also on how early Jews and Christians read it, she says that at creation God created the first covenant, binding all living things in unity with God. Adam was created not so much as a steward but as a high priest, leading all creation in worship of God. Later the temple was modelled after the whole of creation. Sin broke the creation covenant and ultimately brought about the environmental crisis we face now. Jesus died not just to forgive our sins but to restore the whole of the creation covenant. Turning to him involves accepting how we have sinned by destroying creation and gives us hope that by God's grace it can be restored.[20]

Conclusion: faith in action

Loving our worldwide neighbours involves more than telling them about Jesus. James 2:14–17 talks about putting our faith (what we believe in our heart) into action. James says if we don't help people with their physical needs, our kind words are useless.

Would you continue on a path you knew would destroy your house and family? Climate change is destroying our brothers and sisters around the world. God has promised a new heaven and a new earth, but does he want to fill it with people who wrecked the old earth? Would you offer a new house to tenants who trash houses? Destroying the earth so Jesus can return is no better than firing nuclear missiles to bring about the end of the world. How will God judge those of us who take part in ruining the creation he told us to take care of? Revelation 11:18 shows God does not approve of such people:

> The nations were angry,
>   and your wrath has come.
> The time has come for judging the dead,
>   and for rewarding your servants the prophets
> and your people who revere your name,
>   both great and small –
> and for destroying those who destroy the earth.

What other reasons can you think of for caring for creation?

What do you think is the most compelling reason to care for creation?

Creation is not the only thing to care about (see chapter 12). But caring for it is an essential part of our acts of love and witness and worship of the creator. We are called to care for God's world as a way of respecting the creator, caring as God does and obeying God's commands. By doing what we can to care for the **environment**, we can show our love for the vulnerable, who are most affected by climate change, and we can work for justice. We can press governments, powerful organisations and influential people to do what they can too. We can work in unity with other Christians. This can be an act of witness.

What young people are doing

Elsie Luna, 11, heard about the 100 companies responsible for the most **carbon emissions** globally. Many of the companies have offices in London. Elsie went to challenge the leaders of the companies to reduce emissions.[21]

Creator God,
we thank you for making such a beautiful world. It is amazing to see all the colours, hear the birds singing, smell the flowers and feel the snow. What you have made is good. Help us to keep it this way. Help us to care about those who pay the price for our easy lifestyles, and to do what we can to help them. Amen

Loving our neighbours worldwide: Great Lakes Outreach (GLO) in Burundi

greatlakesoutreach.org

GLO

As we mentioned above, Burundi is among the worst sufferers of the effects of climate change. Deforestation and soil erosion limit food production. Food shortages are increased by **droughts** in some regions and floods in others. Burundi is the country rated as the most vulnerable to food shortage, even though its **carbon emissions** are very low. Fewer than two in every 100 in Burundi have electricity in their home.

GLO is a Christian charity in Burundi that partners with projects providing food, education, housing, jobs and Christian teaching.

Young people around the world: **From abandonment to helping those in need in Burundi**

GLO

In Burundi, Grace Nduwimana was abandoned at birth, probably because her mother was unable to provide for her. She was found inside a toilet (not a bathroom, an actual toilet). A local nurse rescued her and cleaned her. When she was five days old, a woman called Chrissie took her in and brought her up. Grace was deaf until she was six months old (possibly due to strong antibiotics she had been given after being found in the toilet). After being prayed for and anointed with oil, her hearing returned.

As a teenager, Grace babysat for the Guillebaud family, who set up GLO. After completing her education, Grace started to work for GLO, helping others who are in need as she once was. Grace is aware of the impact of climate change, and says, 'Unpredictable weather patterns make it difficult to plan for the planting season in Burundi. Many people here are afraid of starving to death.'

Jamie's tips

1. The people who suffer most from the effects of climate change generally live in poor countries, such as Burundi. Pray for these people.

2. Research some of the island nations that are under threat of going under water.

3. Read more Bible passages about creation. Job 39 is a good place to start.

4. God is the creator, and we are made in God's image, so we can be creative too. Create something to celebrate the wonder

of creation, such as a picture, photograph, collage, poem or descriptive paragraph. Or write out and decorate this verse:

> You alone are the Lord. You made the skies and the heavens and all the stars. You made the earth and the seas and everything in them. You preserve them all.
> NEHEMIAH 9:6 (NLT)

5 One person in a rich country like the UK emits as much **carbon dioxide** as hundreds of people in Burundi, which is unfair. Read Part 3 to learn how to reduce your **carbon footprint**, so that there is less inequality between you and the people of Burundi.

6 Jesus said, 'Consider how the wild flowers grow' (Luke 12:27). Think about the flowers, plants and animals and ways of looking after them.

7 Pick up litter. This can be as simple as picking rubbish up in parks in your local community. If you visit a beach, pick up rubbish to stop it polluting the sea. It is best to wear gloves (or use a litter picker) to avoid harming yourself. Recycle what you can.

What young people are doing

Alice Lonsdale, 4, picks up litter with her mother Emily

Glenda Howcroft

8 Find out from people who go to climate strikes or protests, or who take other action against climate change, what their reasons are. This might help you become clearer about your own reasons for taking action.

9 Be creative and make something, such as cakes, biscuits, cards or crafts, to raise money to provide food, education and other assistance in Burundi. To donate, go to greatlakesoutreach.org.

10 Stop polluting the sea. Avoid products containing microbeads, tiny plastic particles that are a growing source of ocean pollution, killing animals. Avoid products with polyethylene or polypropylene on the ingredients list. Choose eco-friendly cleaning products or make your own cleaning products.[22] You can use natural soap nuts in the washing machine instead of washing powder with chemicals in.[23] Vinegar can be used as a water softener and to remove stains.

11 Offer to give a talk to a youth group, church or another group on why we should care about climate change and what we can do. You could use the information from this book to help.

12 Join (or form) a group of people fighting for **climate justice**. Encourage each other, share ideas of actions you can take and work together so that you can achieve more.

Which of these do you already do?

____

____

____

Which will you attempt to do this month?

____

____

____

4 Love in action

Bible passages

1 CORINTHIANS 13:3–8

If I give all I possess to the poor and give over my body to hardship that I may boast, but do not have love, I gain nothing.

Love is patient, love is kind. It does not envy, it does not boast, it is not proud. It does not dishonour others, it is not self-seeking, it is not easily angered, it keeps no record of wrongs. Love does not delight in evil but rejoices with the truth. It always protects, always trusts, always hopes, always perseveres.

Love never fails.

MATTHEW 25:31–40

'When the Son of Man comes in his glory, and all the angels with him, he will sit on his glorious throne. All the nations will be gathered before him, and he will separate the people one from another as a shepherd separates the sheep from the goats. He will put the sheep on his right and the goats on his left.

'Then the King will say to those on his right, "Come, you who are blessed by my Father; take your inheritance, the kingdom prepared for you since the creation of the world. For I was hungry and you gave me something to eat, I was thirsty and you gave me something to drink, I was a stranger and you invited me in, I needed clothes and you clothed me, I was sick and you looked after me, I was in prison and you came to visit me."

'Then the righteous will answer him, "Lord, when did we see you hungry and feed you, or thirsty and give you something to drink? When did we see you a stranger and invite you in, or needing clothes and clothe you? When did we see you sick or in prison and go to visit you?"

'The King will reply, "Truly I tell you, whatever you did for one of the least of these brothers and sisters of mine, you did for me."'

If you were asked to summarise the Bible in two words, which words would you choose?

____

____

There is no single correct answer for this. We might choose 'love' and 'Jesus'.

Love

'God is love. Whoever lives in love lives in God, and God in them' (1 John 4:16). Love is another way to spell God. Jesus said that the most important commandments are to love God and love others; these sum up all the commandments (Matthew 22:36–40).

The word 'love' appears over 680 times in the NIV translation of the Bible. Even where the word is not used, love is often there in the background. One book, the Song of Songs, is entirely a love poem.

The words translated 'love' express more than one idea.[1] Love in the Bible is not a mushy feeling. For example, the Greek word *agape* refers to sacrificial action, as in John 3:16: 'For God so loved the world that he gave his one and only Son.' We are called to put love into action.

1 Corinthians 13 is a beautiful description of love, again using the word *agape*. Love is kind. Love 'does not envy' (or covet) what others have (see chapter 2). Love is 'not self-seeking'. This means love chooses to serve God and one another instead of focusing on what we want (see Philippians 2:4 and the example of Jesus).[2]

Love does not selfishly use everything up. Love does not pollute or trash the world for those who will come after us. Love involves considering other people by denying some of our preferences.

Hints from the Hawkers

As a family, we enjoyed flying to meet people in different cultures. But how does this show love to them, when flying keeps the world warming and harms some of them? Now we travel by train instead, connect with family, friends and colleagues online, and explore new cultures virtually.

Love 'always protects'. What do (or could) you do to help protect other people, out of love?

How we treat the needy

Matthew 25:31–46 is uncomfortable to read. The words suggest people will be judged by whether they acted lovingly. How do we act towards those who are hungry, thirsty, strangers, naked, sick or in prison? Why are people in such conditions today?

> I was hungry, because extreme heat followed by flooding destroyed our crops.
>
> I was thirsty, because the rains no longer fell as they used to.
>
> I was a stranger, because the rising sea level meant we had to leave our island and travel as refugees to a foreign place.
>
> I needed clothes, because all my possessions were destroyed in a hurricane.

Nuevas Esperanzas

Flooded road after Hurricane Felix

> I was sick, because hotter weather and more rain brought mosquitoes to my region and I caught malaria.
>
> I was in prison, because I joined a climate change protest.[3]

Are you contributing to the climate change that leaves some hungry, thirsty, refugees, without possessions, sick or in prison? Are you ignoring others and leaving them to suffer? Or are you addressing the environmental problems which cause suffering for millions of people?

The rich man and Lazarus

Do you feel uncomfortable with Matthew 25? What about Jesus' story about a rich man and a beggar (Lazarus) who lived at his gate (Luke 16:19–31)? This story is rarely discussed, perhaps because we feel so uncomfortable with it. Jesus did not say that the rich man had forced the beggar into poverty, insulted him or had him beaten and dragged away. The beggar was just there, perhaps being ignored. The next thing we know, the rich man died and went to hell, while the beggar died and was at Abraham's side. The implication is that the rich man was punished for not helping the beggar.

Ouch! How many suffering people do we ignore daily? And how regularly do we add to their suffering, through our own thoughtless actions?

When the rich man asks if Lazarus can be sent to warn his family, so that they will not also be sent to a place of torment, the reply is, 'They will not be convinced even if someone rises from the dead' (v. 31). There seem to be some people who ignore evidence, whether it comes from science or experience.

The golden rule of love

Jesus taught, 'In everything, do to others what you would have them do to you, for this sums up the Law and the Prophets' (Matthew 7:12). This is known as the 'golden rule of love', or simply the golden rule.

What could the golden rule mean about how to treat the people who will be most affected by climate change?

Does the golden rule apply just to people we know? No. It's about people everywhere and in every time. We are all God's children (Malachi 2:10). We have a duty to future generations, because our actions affect them (Exodus 34:7; Numbers 14:18).

Jostein Gaarder, author of bestseller *Sophie's World*, has said:

> You shall do to the next generation what you wished the previous generation had done to you… Based on the principle of reciprocity, we should only permit ourselves to use non-renewable resources to the extent that we at the same time pave the way for our descendants to be able to manage without the same resources.[4]

People in the future will suffer more from climate change than us. We are using up resources that cannot be replaced.

What would it look like to do what Jostein Gaarder said? Or to create **carbon emissions** only to the extent that we offset them (for example, by helping to save forests and investing in new technologies to reduce carbon)?

What could I do to show that I want to treat the next generation how I want to be treated, when it comes to the climate crisis?

> **Greta Thunberg said:**
> **'If I have children... Maybe they will ask why you didn't do anything, while there was still time to act. You say that you love your children above everything else. And yet you are stealing their future... We cannot solve a crisis without treating it as a crisis.'**[5]

Jesus told his disciples to make disciples of all nations, and he taught them to care for the needy. Our loving actions show we are learning from him. Jesus said, 'By this everyone will know that you are my disciples, if you love one another' (John 13:35).

God of the poor,
thank you for making us your children.
Teach us how we should love others, both near and far,
now and in the future. Thank you, loving God. Amen

Loving our neighbours worldwide: Climate Stewards

climatestewards.org

Climate Stewards is a charity which teaches that we are all called to be good stewards of God's earth, and that voluntary **carbon offsetting** is a great way to do this.

Climate Stewards partners with projects in several countries, aiming to reduce **carbon emissions** in a **sustainable** way which is 'owned' by local people. This includes tree planting; providing water filters (so people can make water safe to drink without burning charcoal); and installing fuel-efficient (wood-saving) smokeless stoves. All three reduce deforestation and air pollution.

Dr Alex Zahnd/Climate Stewards partner, RIDS-Nepal

Dr Alex Zahnd/Climate Stewards partner, RIDS-Nepal

Above: **Open fire fills a house with smoke**

Right: **After receiving a smokeless stove**

People in islands like Samoa urgently need **carbon emissions** reduced. Samoa's prime minister said climate change is an 'existential threat' to island nations. The islands may cease to exist, as rising sea levels cover them and drive people from their homes.[6]

Young people around the world: Extreme weather conditions in Samoa

A 17-year-old in Samoa says, 'Climate change is a day-to-day thing for people in Samoa. As a teenager growing up, it is my reality to see cyclones, flash floods, general flooding in the area we live in, **droughts** and all sorts of extreme weather we see every day… My wish for the world is that everyone can concentrate on [limiting the temperature rise to 1.5 degrees] because that is the only way our Pacific islands can survive.'[7]

Jamie's tips

1. If you are a Christian and have chosen to follow God, don't forget to pray regularly. Pray for the world.

2. Read 1 Corinthians 13:4–7, but replace the word 'love' or 'it' with your own name. See how it sounds. Is it true now? What would you have to change to make it true?

3. Think about things you love about our planet.

4. Romans 13:10 says, 'Love does no harm to a neighbour.' List anything that you do which might cause harm to someone else. Then list something you plan to do to reduce this harm.

5. Study these Bible passages about love: Proverbs 19:17; Colossians 3:12–14; Titus 3:14; 2 Peter 1:5–7; 1 John 3:17–18; 1 John 4:20–21; 2 John 5–6; Jude 2.

6. Show that you love God, who created the world, by doing something this week to care for the world.

7. Use the carbon calculator at climatestewards.org/offset to estimate your **carbon emissions** for the past year. Work out ways to reduce this.

8. Following on from number 7, offset your emissions by donating money to the projects run by Climate Stewards.

9 Go a step further by holding a fundraising event with your church or another group to raise money to offset the **carbon emissions** of those in the group.

10 Think about what you would want others to do for you if you lived in a place like Samoa, which has been terribly affected by climate change. What will you do to keep emissions down, to try to keep these islands above sea level?

____

____

____

11 C.T. Studd (cricketer and missionary) said, 'If Jesus Christ be Lord and died for me, then no sacrifice can be too great for me to make for him.' Think of something which you are willing to sacrifice to help care for God's creation.

12 Get to know someone from another country that is affected by climate change. Ask what the situation is like there. You may find that you care about climate change more if you have a friend who is in an affected country.

Which of these do you already do?

____

____

____

Which will you attempt to do this month?

____

____

____

5 Hope in action

Bible passages

JONAH 3:1—4:2

Then the word of the Lord came to Jonah a second time: 'Go to the great city of Nineveh and proclaim to it the message I give you.'

Jonah obeyed the word of the Lord and went to Nineveh. Now Nineveh was a very large city; it took three days to go through it. Jonah began by going a day's journey into the city, proclaiming, 'Forty more days and Nineveh will be overthrown.' The Ninevites believed God. A fast was proclaimed, and all of them, from the greatest to the least, put on sackcloth.

When Jonah's warning reached the king of Nineveh, he rose from his throne, took off his royal robes, covered himself with sackcloth and sat down in the dust. This is the proclamation he issued in Nineveh:

> 'By the decree of the king and his nobles:
>
> Do not let people or animals, herds or flocks, taste anything; do not let them eat or drink. But let people and animals be covered with sackcloth. Let everyone call urgently on God. Let them give up their evil ways and their violence. Who knows? God may yet relent and with compassion turn from his fierce anger so that we will not perish.'

When God saw what they did and how they turned from their evil ways, he relented and did not bring on them the destruction he had threatened.

But to Jonah this seemed very wrong, and he became angry. He prayed to the Lord, 'Isn't this what I said, Lord, when I was still at home? That is what I tried to forestall by fleeing to Tarshish. I knew that you are a gracious and compassionate God, slow to anger and abounding in love, a God who relents from sending calamity.'

2 CHRONICLES 7:13–14

'When I shut up the heavens so that there is no rain, or command locusts to devour the land or send a plague among my people, if my people, who are called by my name, will humble themselves and pray and seek my face and turn from their wicked ways, then I will hear from heaven, and I will forgive their sin and will heal their land.'

Hope in the Bible

'And now these three remain: faith, hope and love. But the greatest of these is love' (1 Corinthians 13:13). Faith, hope and love are a wonderful trio. To live well, we need them all. Each feeds into the others. We have already discussed faith and love in action. This chapter is about hope.

Do you feel more hopeful or hopeless when you think about climate change? Why?

____

____

____

____

____

____

Thinking about climate change and other big issues can leave us feeling hopeless and discouraged. Figure 2 in chapter 2 (p. 36) looks bleak. Is it too late to stop disastrous climate change? Is adapting to it all we can do?[1] If we feel hopeless, we are unlikely to do anything to try to change the situation, because we believe that our efforts will be pointless. We need hope or we will not act. Thankfully, God is a God of hope (Romans 15:13).

Did you know?

At least 16 words in the Bible are translated 'hope'. Taken together, these words occur about 130 times in the Bible. In the Psalms and the prophets, 'hope' is often a verb – an action word. Hope and action go together.[2]

As Christians, we have big-picture hope, ultimate hope, when we take an eternal perspective – that is, the perspective of 'the final achievement of all God's purposes for his creation when he brings this temporal history to its end and takes the whole creation, redeemed and renewed, into his own eternal life'.[3] This is the time when 'the creation itself will be liberated from its bondage to decay' (Romans 8:21).

We also have small-picture hope that positive change will happen in this life. In the Old Testament economy, debts were cancelled and the land rested every seven years. Every 50 years, slaves were set free and land that had been sold was returned to its owners. No hardship lasted forever. What about hardship facing people now? If we are faithful to God and love our neighbours, we hope God may save us from environmental disaster. In the next chapter we will see how Joseph helped prevent a disaster and how Noah helped his family remain safe. Here, we will think about Jonah.

A second chance

Jonah went to Nineveh (in modern-day Mosul, Iraq), a city known at the time for its wickedness, including inflicting terrible suffering on others (see Nahum 3). Jonah warned that the city would be destroyed in 40 days (Jonah 3:4). Previously Jonah had disobeyed God and run off in the opposite direction instead of proclaiming God's message (Jonah 1:3). God gave Jonah a second chance. After the king led the people of Nineveh to repent, God gave *them* a second chance as well. God is known for giving second (and third, and more) chances – to Peter (John 21:15–17), Moses (Exodus 2—3), David (2 Samuel 11—12), Samson (Judges 16; Hebrews 11:32–34), Abraham (Genesis 16:4; 20:2) and, indeed, to all of us.

Jonah was angry that God gave Nineveh another chance. But he knew that God is 'a gracious and compassionate God, slow to anger and abounding in love, a God who relents from sending calamity' (Jonah 4:2). Let us pray that God will again relent from sending calamity in our time.

Nineveh heeded a warning

- What would it take today to cause leaders and citizens to heed warnings about climate change and turn away from the path to destruction?

Turning to God

When King Solomon had just dedicated the new temple for worship, God made him a promise (2 Chronicles 7:13–14): in a **drought** or a swarm of locusts or widespread disease, if God's people would be humble and pray and seek God's face and turn from wicked ways, God would forgive them and heal their land. As we wrote this in 2020, the world had just experienced all of these – **drought**, locust swarms and a pandemic (as well as forest fires, floods and cyclones). All of these can be caused or made worse by climate change.

These verses are a specific promise to the people of Israel at a particular time, and we should see them in context. The whole nation was being called to pray. In the context of worship, God's people were encouraged to turn to God whenever the **environment** was in danger. In an environmental crisis, we can still pray, seek God and turn from wrong ways of living.

Sachira Kawinda on Unsplash

Dry soil in a drought

Is there anything you choose to turn away from?

___

___

___

Praying is not just speaking. Praying includes asking God how we should act, listening and following God's prompting.

Burning bushes

VanderWolf/stock.adobe.com

God spoke to Moses through a burning bush. Bushes are still burning in our day. Hot, dry weather as a result of climate change causes widespread wildfires – in Australia, the Amazon, Siberia, Indonesia, sub-Saharan Africa, America's west coast and many other lands, including the UK.

God spoke to Moses through the burning bush. Stop and look at an image of a forest fire. Ask what God wants to say to you through this. Is there an action God is asking you to take? Make time to pray.

Hope leads to action

Hope can motivate us to act. If you hope to pass exams, you study. If you hope to get a job, you send good application forms. If we hope to reduce the damage done by climate change, we need to do our part. Shane Lopez said, 'Hope is created moment by moment through our deliberate choices.'[4]

C.S. Lewis on hope

Hope is one of the Theological virtues. This means that a continual looking forward to the eternal world is not (as some modern people think) a form of escapism or wishful thinking, but one of the things a Christian is meant to do. It does not mean that we are to leave the present world as it is. If you read history you will find that the Christians who did most for the present world were just those who thought most of the next... [for example] the English Evangelicals who abolished the Slave Trade, all left their mark on Earth, precisely because their minds were occupied with Heaven.[5]

In what ways have you put hope into action?

____

____

____

____

We can join with many others to work for justice and change the arc of history.

Cherice Bock on communal hope

I can convince myself that things will probably be fine in my lifetime... But if I think of hope in a long-term sense, in the sense of the arc of history bending toward justice, I see history from a different vantage point. I recognize my contribution to the history of justice and righteousness, and I want to be part of creating that community of shalom, rather than only creating a momentary blip of personal comfort in my own lifetime. This biblical vision of communal hope rather than individual goal completion offers a broader and deeper purpose for acting on the part of the planet and all creation to steward God's created world with care, looking toward the future with hope for the entire community of creation.[6]

Not fake news

Jeremiah and Zephaniah described false prophets who brought fake news, promising hope without catastrophe when that was not what God was saying (Jeremiah 28:15–16; Zephaniah 3:4). The true prophets sent by God offered hope beyond catastrophe: hope that God would be with them even when life was hard. They shouted warnings that people needed to change. We still need true prophets to shout warnings today.

Difficult times will come. God promises to be with us in the difficulties, not to take us out of them (Psalm 23:4). We must integrate our active hope (doing what we can) with our trusting hope that God will be with us even in difficult times, and ultimately the big-picture hope that God offers eternal life and restored creation.

Feed hope

It is said that a Cherokee grandfather was teaching about life. 'There are two wolves inside me, and they are always fighting,' he said. 'One is greed, darkness and despair, the other is kindness, light and hope. Which wolf wins? The one you feed.'[7]

We can choose to feed hope by connecting with others who have hope and acting to demonstrate our hope in a better future.

Hope for the environment

Despite the damage already done to the **environment**, it is not too late for us to slow down further harm and reduce the damage. What we do now can make a positive difference. There is hope.

In 2020 during lockdown, pollution and **carbon emissions** fell,[8] as planes stopped flying, people stopped commuting and roads emptied of traffic. Canals in Venice cleared, and birds could be heard singing where usually only traffic noise was heard. This proves that we can reduce emissions if we drastically reduce flights and car travel. It is not too late yet.

'Here come the young,' sang Martyn Joseph, hopefully.[9] Throughout this book we see young people who are positively acting for change. New technologies are being produced to reduce **carbon emissions**. As you read on, consider which example (or young person or charity) from this book has most struck you as a sign of hope.

Markus Winkler on Unsplash

Cultivating fertile land

What signs of hope are you aware of?

What young people are doing

After hearing a church service about being led by the Holy Spirit, 15-year-old Josiah Barron felt led to get involved with school strikes for the climate. Since then he has been leading the climate strikes in his local area. These have grown from 60 to 1,500 people. As a deputy member of Youth Parliament, Josiah has opportunities to speak with MPs about environmental issues.

Josiah says that he is motivated to act by his faith and by hope. He believes that it is important not to focus only on the negatives. He says, 'We're taking steps, the world is changing.' For example, he has helped his church to start meeting targets to become an Eco Church (see ecochurch.arocha.org.uk). The members of the youth group and other church clubs now car-share. He also met with his local council, who are looking into setting up a museum to show the impact of climate change, to challenge people to act.

Josiah Barron

Josiah says, 'Meeting other people at events gives hope and helps us all to continue.'

Prayer

God of hope,
help us put our hope into action. Show us what you want us to do, and give us the courage to act. In Jesus' name, Amen

Did you know?

Bangladesh is home to over 160 million people. Two thirds of it is less than five metres above sea level, and almost 80% of the land is prone to flooding. By 2050 it may lose 11% of its land, affecting 15 million people. Other impacts include cyclones, storm surges, extreme temperature and drought. The Global Climate Risk Index rated Bangladesh in the top ten of countries most at risk from climate change (2000–2019).[10]

Loving our neighbours worldwide: Baptist Missionary Society (BMS) in East Asia

bmsworldmission.org

> **BMS on climate change:**
> **'At BMS, we are convinced that addressing climate change is a matter of missiological urgency.'[11]**

BMS has workers in some of the most fragile places on earth. They work in areas of relief, development, education, health, justice, leadership and church ministries. BMS has a policy on creation stewardship, which other charities and churches could learn from.

BMS worker Jack (we've changed his name for security reasons) set up a consultancy company in East Asia, to help garment factories reduce their **carbon footprint** by up to 40%. Jack has taught factory owners to optimise their use of dye chemicals, leading to less pollution in rivers.

Young people around the world: A school flooded by monsoon rains in Bangladesh

Photo and details provided by BMS

16-year-old Surbana attends the Baptist Missionary Integrated School in Bangladesh. This school was founded by a BMS missionary, to cater for blind pupils and other vulnerable children. From the age of 11, blind and sighted children are integrated together.

Until recently, 2–3 feet of monsoon waters flooded the school every year. The school's boundary wall collapsed as a result. The flooded bathrooms were unsanitary. Now the wall has been replaced and reinforced with steel. New, clean bathrooms have been built.

Education, especially for girls, provides hope. Surbana, full of hope, declares, 'Our school gives us a way to fulfil our dreams.' She would like to become a lawyer.

Jamie's tips

Zac Wolff on Unsplash

1 Feed hope by looking at this photo from Milan in Italy, where two tower blocks contain about 800 trees, and over 15,000 shrubs and other plants. These buildings use renewable energy from solar panels. Several other cities have, or are planning, similar vertical forests.

2. List Bible verses which mention hope. If you can't think of any, look up Ezra 10:2; Isaiah 40:31; Lamentations 3:21–25; Hosea 2:15; Romans 5:1–2; 1 Thessalonians 4:13; 1 Peter 3:15.

3. Look at the BMS website to see stories of hope from around the world, and download a free magazine on climate change.[12]

4. Read slowly Jeremiah 29:11:

   'For I know the plans I have for you,' declares the Lord, 'plans to prosper you and not to harm you, plans to give you hope and a future.'

   Pray about this verse. Ask God to reveal what you should do to show hope in action, as part of the plans that God has for you. Spend time in quietness reflecting on this.

5. Plant a tree or a plant, to demonstrate that you have hope that it will grow.

6. BMS gives hope to those in need – providing immediate practical help and also sharing the gospel (ultimate hope). Be part of this by sending BMS a donation and praying for their work. Perhaps hold a quiz night with questions about climate change and request donations for taking part. You could do this online if you want people in different places to take part.

7. Do you have hope that your city or local area can become **carbon neutral**? Feed hope by reading about what other regions have done.[13] Contact your local council and ask them to reduce their **carbon emissions**, following the examples of others.

8. One way to build hope is to link up with other people and find out what positive acts are already going on. Find out what other people are doing, by word of mouth, social media, an internet search or connecting with an environmental group (Appendix C).

9 Greta Thunberg said, 'Hope is something you have to earn', by your actions.[14] Write down three ways in which you will put your hope into action this month.

____

____

____

10 Children and young people often have contagious hope. Speak with young people who are interested in **climate justice** and ask them for ideas about what you could do. If you don't know any young activists, read examples from this book or look up Greta Thunberg on the internet for ideas. The Methodist Church has set up gatherings for young people who are passionate about **climate justice** ('Green Agents of Change'). Could your church do something similar (not necessarily just for young people)?

11 Follow the example of 15-year-old Josiah by helping your church (or school, university, group or workplace) move towards being more 'eco'. For ideas, see ecochurch.arocha.org.uk and the church **carbon footprint** calculator at 360carbon.org.

12 Share hope with others. Tell someone what people are doing to reduce **carbon emissions.** Share examples from young people in this book.

Which of these do you already do?

____

____

Which will you attempt to do this month?

____

____

6 Wisdom in action

Bible passages

GENESIS 6:13–14, 17–22

So God said to Noah, 'I am going to put an end to all people, for the earth is filled with violence because of them. I am surely going to destroy both them and the earth. So make yourself an ark of cypress wood; make rooms in it and coat it with pitch inside and out… I am going to bring floodwaters on the earth to destroy all life under the heavens, every creature that has the breath of life in it. Everything on earth will perish. But I will establish my covenant with you, and you will enter the ark – you and your sons and your wife and your sons' wives with you. You are to bring into the ark two of all living creatures, male and female, to keep them alive with you. Two of every kind of bird, of every kind of animal and of every kind of creature that moves along the ground will come to you to be kept alive. You are to take every kind of food that is to be eaten and store it away as food for you and for them.'

Noah did everything just as God commanded him.

GENESIS 41:28–36

'It is just as I said to Pharaoh: God has shown Pharaoh what he is about to do. Seven years of great abundance are coming throughout the land of Egypt, but seven years of famine will follow them. Then all the abundance in Egypt will be forgotten, and the famine will ravage the land. The abundance in the land will not be remembered, because the famine that follows it will be so severe. The reason the dream was given to Pharaoh in two forms is that the matter has been firmly decided by God, and God will do it soon.

'And now let Pharaoh look for a discerning and wise man and put him in charge of the land of Egypt. Let Pharaoh appoint commissioners over the land to take a fifth of the harvest of Egypt during the seven years of abundance. They should collect all the food of these good years that are coming and store up the grain under the authority of Pharaoh, to be kept in the cities for food. This food should be held in reserve for the country, to be used during the seven years of famine that will come upon Egypt, so that the country may not be ruined by the famine.'

PROVERBS 22:3

Sensible people will see trouble coming and avoid it, but an unthinking person will walk right into it and regret it later. (GNT)

Wisdom

The Bible says wisdom was involved in creation (Proverbs 3:19), and it calls wisdom a life-giving tree or fountain (Proverbs 3:18; 13:14). Prophets said that when people abandon wisdom, environmental crisis follows: 'There will be no grapes... no figs... their leaves will wither' (Jeremiah 8:13). Noah, Joseph and others wisely took action to prevent problems, including danger related to the weather or climate.

Noah

God told Noah to build an ark to protect himself and his family from flooding. God also commanded Noah to take a male and a female of every kind of bird and animal on to the ark, so they could reproduce. God cared about every species and did not want any to become extinct.

Although Noah might have been ridiculed for building a huge boat in a dry place, he obeyed God, not being influenced by what others might think.

Karson on Unsplash

Have you ever been ridiculed for obeying God?
Would you be willing to take that risk?

Other people ignored Noah and were destroyed. Still today, many people are ignoring warnings. After the flood, God put a rainbow in the sky as a sign that he would not flood the whole earth again (Genesis 9:13–17).

Joseph

In Genesis 41, God revealed to Joseph (through Pharaoh's dream) that seven years of plentiful harvests would be followed by seven years of famine, when crops would fail. Though only 30 years old (v. 46), Joseph was wise and he planned ahead. During the years of plenty, he made sure that food was stored away so it could be used during the famine. This was not a selfish act. During the famine, the store fed not only Egypt but also people from surrounding lands (v. 57). After the famine, Joseph gave out seeds so that people could start to grow food again (Genesis 47:23).

In the years of plenty, people ate less than they could, to save food for the years that would follow. What sacrifice could you make now to help reduce the suffering of others in the years ahead?

Noah and Joseph both showed wisdom and integrity (despite their mistakes).

The wise see danger coming and avoid it

Proverbs 22:3 tells us that sensible (or wise) people anticipate trouble and act to avoid it. This warning is worth repeating (in Proverbs 27:12). The writer suggests that it is foolish to carry on daily life with no concern about forthcoming danger.

Logan Weaver on Unsplash

During the Covid-19 pandemic, wise people heeded warnings and changed their behaviour significantly to help save lives. Climate change may seem further off, but it risks far more lives.

What young people are doing

Swedish activist Greta Thunberg started to strike from school when she was 15, sitting outside the Swedish parliament to raise awareness of the climate emergency. This schoolgirl has inspired people all over the world to take action for **climate justice**. She does not let the fact that she is on the autistic spectrum stop her, but rather she uses the way her brain is wired as a strength.

During the Covid-19 pandemic, Greta said, 'If one virus can wipe out the entire economy in a matter of weeks and shut down societies, then that is a proof that our societies are not very resilient. It also shows that once we are in an emergency, we can act and we can change our behaviour quickly.'[1]

Which of our behaviours would it be wise to change in response to the climate emergency?

Did you know?

You can follow Greta at facebook.com/gretathunbergsweden, instagram.com/gretathunberg and twitter.com/GretaThunberg.

Spencer and White sound a warning:

> To continue our present reckless experiment with the Earth's future climate would be a morally gross failure both towards those who are less fortunate than ourselves in their material resources and towards future generations… the sheer scale and impact of climate change makes it a priority… Global warming demands a response.[2]

Did you know?

At least 8.8 million people are thought to die every year from air pollution. That is about 1,000 people every hour.[3]

Asking for wisdom

God told Solomon he could ask for anything he wanted. Solomon asked for wisdom (1 Kings 3:5–12).

What would you ask for?

____

____

____

'The fear of the Lord is the beginning of wisdom' (Proverbs 9:10). Ask God for wisdom (James 1:5). Studying and applying the Bible helps us to grow in wisdom. Wisdom is not the same as knowledge; it's shown in our behaviour: 'Who is wise and understanding among you? Let them show it by their good life, by deeds done in the humility that comes from wisdom' (James 3:13).

How can we act wisely now? In this climate emergency, we can contribute to the problem or to the solution.

The following chapters will help us better understand what sort of good deeds we can do as we seek to be wise in our environmental actions. As Jesus said, 'Whoever has ears, let them hear' (Matthew 11:15).

God our protector,
please give us wisdom to see when there is trouble ahead and take action to reduce the risk. Help us to be wise in how we look after the planet and reduce some of the problems caused by climate change. Help us make wise choices that will protect ourselves and others. Amen

Did you know?

Climate change can increase the risk of earthquakes and volcanic eruptions.

The changing climate can cause a rise in the frequency of the most destructive typhoons or hurricanes. It is believed that these reduce pressure in the atmosphere, which leads faults in the earth to move more easily, causing earthquakes.

Flooding can also trigger earthquakes as erosion and landslides reduce the weight on tectonic plates, which then move more easily.

Heavy rain and rising sea levels may also trigger volcanoes to erupt.[4]

Loving our neighbours worldwide: International Nepal Fellowship (INF)

inf.org

INF works in the western region of Nepal, a country vulnerable to the effects of climate change, including flooding, landslides, earthquakes, **drought**, fire and soil erosion. The Global Climate Risk Index rated Nepal the country fourth most vulnerable to climate risk in 2017.[5] An earthquake in Nepal in 2015 killed nearly 9,000 people and injured nearly 22,000.

INF Nepal/Rautahat Flood Response 2019

Flood damage in Nepal, 2019

As well as providing health care, INF provides emergency relief and recovery work after disasters and helps reduce risk by building communities' resilience to disaster. INF set up self-help groups to educate communities about disasters and the effects of climate change, helping them to reduce the risk and impact of disasters. These groups have taken collective action in a range of ways – for example, planting trees to reduce the risk of soil erosion.

Young people around the world:
An earthquake and mudslides in Nepal

INF Nepal

Bhim Maya Gurung, 16, is from Laprak, one of the villages which was hit badly by the Nepal earthquake in 2015. Her house completely collapsed during the earthquake. She and her parents and two brothers moved into a temporary shelter. She says, 'My village is listed as one of the prone areas for mudslides. We live in fear each day, especially during the monsoon.'

Bhim Maya has poor eyesight. In the small shelter, she often bumped into the walls and fell to the ground. She would spill water and scatter grains by accident, and struggled to help her mother with chores. She needed friends to help her find her way to school. She could not read what teachers wrote on the board. She felt depressed and decided to give up on her studies.

After the earthquake, INF started a project in Bhim Maya's region. The staff inspected her eyes and consulted with her parents to have her sent for an eye check-up. She was prescribed glasses and says, 'I was overjoyed to see things so clearly after so long in my life. It was like a miracle to see the beautiful places and things on my way back to my village.' She has decided to continue her studies. 'It feels like having a new life again when there was no hope,' she says. She wants to be a teacher for children with disabilities.

Jamie's tips

1 Learn more by watching a three-minute video about INF's Community Resilience Project, at **inf.org/news/response-recovery-resilience**.

2 Pray that God will give you wisdom about what *you* should do to help reduce problems related to climate change. Wait, listen and write down any ideas which come to mind.

3 Write out and try to memorise Proverbs 22:3.

4 Think of other Bible passages about taking action to prevent disaster. For example, the prophets warned people to change and told of dangers to come. See also Matthew 7:24–27.

5 During the Covid-19 pandemic, we were told to 'stay home and save lives'. People put rainbows in their windows as signs of hope (remember Noah's rainbow?). Create a motto or a picture about climate change. Try making a banner or poster with this on, and perhaps put it in a window or take it on a march.

6 Keep studying the rest of this book, so that you find out more about climate change and the actions you can take to help reduce the risk or prepare for the consequences.

7 Who could you warn about climate change and the need to act now? Could you give or lend a copy of this book to a friend or family member? Or ask for climate change to be addressed in a church service, youth meeting, lesson or a talk at the place where you study or work?

What Jamie did

I wrote a letter asking short-term mission organisations to provide flight-free mission opportunities instead of sending all their gap-year/short-term volunteers on flights. There could be opportunities to volunteer within the UK or to hear about projects online instead of flying to them, or to travel by train and boat instead of by air. Premier Radio interviewed me about this. We have received replies from a few mission agencies. Serving In Mission, Baptist Missionary Society and European Christian Mission were especially positive and agreed to make changes. Global Connections decided to consider the topic of climate change at a conference.

8 It took Noah a long time to build a huge ark. Joseph spent seven years storing food before the famine. For us, too, acting to reduce problems caused by climate change will take time. Put a reminder on your phone, calendar or diary for the first day of every month this year to reflect on how you are getting on with reducing your **carbon footprint** and taking action on **climate justice**.

9 Raise funds for INF's resilience-building work. Could you put on a movie night with popcorn and ask for donations? The movie *Paddington* (2014) starts with an earthquake so that might be appropriate. Or try *WALL-E* (2008) or *The Lorax* (2012), both fun family films that make important points about climate change. Alternatively, how about a board-game evening with snacks available?

10 Choose one thing that you are willing to give up to help save the lives of your grandchildren or the next generation. What will it be?

11 Raise awareness that we need to take the climate emergency as seriously as the Covid-19 crisis. Write to a newspaper, politician or organisation to get this point across. As Edward Bulwer-Lytton wrote, 'The pen is mightier than the sword.'

12 Many people prepare for risks by saving money. Do you have any savings? If so, have you considered switching to an ethical bank? If you do not have a bank or building society account yourself, ask which one your church, denomination or group uses, and whether it is an ethical one.

Did you know?

Some banks are known as 'ethical banks'. The website beemoneysavvy.com defines these as 'banks whose business model is designed around making sure they have no negative impact on the environment or on society'.

Triodos Bank is one example of an ethical bank that positively chooses to invest in sustainable energy.[6]

Which of these do you already do?

______

______

______

Which will you attempt to do this month?

______

______

______

PART 3

What we can do

7 Travel: the compassion to change a journey

Bible passage

LUKE 10:25–37

On one occasion an expert in the law stood up to test Jesus. 'Teacher,' he asked, 'what must I do to inherit eternal life?'

'What is written in the Law?' he replied. 'How do you read it?'

He answered, '"Love the Lord your God with all your heart and with all your soul and with all your strength and with all your mind"; and, "Love your neighbour as yourself."'

'You have answered correctly,' Jesus replied. 'Do this and you will live.'

But he wanted to justify himself, so he asked Jesus, 'And who is my neighbour?'

In reply Jesus said: 'A man was going down from Jerusalem to Jericho, when he was attacked by robbers. They stripped him of his clothes, beat him and went away, leaving him half-dead. A priest happened to be going down the same road, and when he saw the man, he passed by on the other side. So too, a Levite, when he came to the place and saw him, passed by on the other side. But a Samaritan, as he travelled, came where the man was; and when he saw him, he took pity on him. He went to him and bandaged his wounds, pouring on oil and wine. Then he put the man on his own donkey, brought him to an inn and took care of him. The next day he took out two denarii and gave them to the innkeeper. "Look after him," he said, "and when I return, I will reimburse you for any extra expense you may have."

'Which of these three do you think was a neighbour to the man who fell into the hands of robbers?'

The expert in the law replied, 'The one who had mercy on him.'

Jesus told him, 'Go and do likewise.'

The good Samaritan

The good Samaritan is one of Jesus' best-known parables. Because of it, someone who goes out of their way to help a stranger is now called a good Samaritan. Samaritans is a charity offering free 24-hour help to people in emotional distress. Charities and hospitals are named after the good Samaritan. Many people think that 'Samaritan' implies a good person.

Jesus told this story to a religious expert who saw Samaritans differently. Jews and Samaritans hated each other. Jews had destroyed the Samaritan temple on Mount Gerizim, and Samaritans had put human bones in the holy Jewish temple to show disrespect and make it unclean.

Shortly before Jesus told this parable, he and his disciples had been turned away from a Samaritan village. Jesus' disciples James and John wanted fire from heaven to destroy the Samaritans (Luke 9:51–56). That's how Jews saw Samaritans then.

The religious expert asked Jesus what he needed to do to inherit eternal life. On being told to love God and neighbour, he replied, 'Who is my neighbour?'

By replying with this story, Jesus showed that we are all neighbours, even Jews and Samaritans. We should help anyone who is in need – even a sworn enemy.

Jesus described the priest and the Levite failing to come to the man's aid, while the Samaritan stopped and helped. Priests were highly respected for their work in the temple. Perhaps the priest thought the man was dead, or not

Jewish. Either way, touching the injured man could make the priest unclean, so he did not get involved. Levites assisted priests in the temple. Perhaps the Levite saw the priest and thought, 'If he went past the injured man, I can do the same.'

Do you ever stop and take positive action, even if other people are not helping (e.g. with climate justice)? Or do you copy those who ignore problems?

The robbers were violent. The priest and the Levite were not violent, but they hurt the man by neglecting him, ignoring his suffering and failing to help him. Proverbs 3:27 states, 'Do not withhold good from those to whom it is due, when it is in your power to act.' Likewise, James 4:17 says, 'If anyone, then, knows the good they ought to do and doesn't do it, it is sin for them.'

Have you ever hurt anyone by not doing good?

The good Samaritan was generous. He interrupted his journey. Although the injured man was a Jew, he tended his wounds with oil and wine. Compassion led the Samaritan to change his journey plans and put an injured enemy on his donkey; perhaps the Samaritan then had to walk alongside. He took the wounded man to an inn and paid two days' wages for the innkeeper to look after the man. He went even further by promising to pay more later if it was needed. This was like leaving his credit card, as he did not know what the innkeeper might charge him afterwards.

> **It's our deeds that matter, not our biblical knowledge:**
> **When Debbie was a child, she was part of a team that took part in an annual Bible quiz, competing against other Sunday schools. One round of the quiz included questions about the parable of the good Samaritan. Dozens of excited children and their parents crowded into the quiz room to take part and see who would win the trophy. Afterwards, Debbie's mother pointed out, 'Everyone walked past a drunk man lying on the pavement to get to the quiz, and no one stopped to help.' What is the point of memorising Bible passages if it does not change how we act?**

Jesus told the religious expert to 'go and do likewise' (v. 37). His listeners would have been shocked to hear the instruction to imitate a Samaritan (the enemy) and not the priests and Levites. The message is to copy people who show love and compassion, not those who don't.

Who should I imitate in the climate emergency?

A member of the group Christian Climate Action said, 'God called the church to take action on climate change, but they didn't. So God called Extinction Rebellion.' Saying this to Christians might be like saying to Jews in Jesus' day that God had used a Samaritan to help an injured man.

Christian Climate Action

Extinction Rebellion (XR) is an international environmental movement established in 2018, aiming to use non-violent civil disobedience to bring about government action to reduce **carbon emissions**. XR protesters have glued themselves to buildings and even a plane, as well as blocking roads. Many are willing to be arrested to make people and governments realise we need to deal with the climate emergency. In many cases in the UK, charges against protestors have been dropped.

Christian groups such as Christian Climate Action work alongside XR. There are also many other groups taking action for the climate (see Appendix C). Some people think that XR goes too far.[1] XR argues that everything else has already been tried, and only protests that inconvenience others work. Who should we imitate? Those who do nothing? Those who cause harm by neglect? Or those who are taking effective action? The parable of the good Samaritan suggests that we should copy compassionate actions, whether or not the person involved shares our faith. We are called to obey God, not people (Acts 5:29).

Who is my neighbour in the climate emergency?

Over the years, there have been many modern adaptations of the parable of the good Samaritan. We have heard it told as the story of the good punk-rocker, the good illegal immigrant and the good Gypsy. What might the parable of the good Samaritan look like in the current climate crisis?

> **A man was afraid of losing his house, his animals and his family, owing to floods as the climate was changing. His food supply was already inadequate, because crops had failed, and he and his family were malnourished and hungry.**
>
> **A missionary flew overhead to her third conference that month, saying, 'It's not my calling to help here.'**
>
> **A church worker drove past, preaching, 'Let the earth be destroyed, and then Jesus will return.'**
>
> **A teenager from another faith saw the man and had compassion on him and his family. She changed her lifestyle to reduce her carbon footprint and persuaded her parents to let the man and his family stay with them until they could find a safer place to live. She raised money for a charity to help the man and his community, providing emergency food, as well as seeds, tools and information about food they could grow in the changing conditions. And she protested and campaigned to get the government to change policies.**

Is this shocking? Do you feel offended? So did the people listening to Jesus. Priests and Levites were respected religious workers, like missionaries and church workers today. Before anyone objects, we should explain that we (Debbie and David) are ourselves missionaries and church workers, and we have told this story against ourselves and not anybody else. For too long we carried on with our own ministries, travel and comfortable lifestyle while ignoring those who were suffering because of the **carbon emissions** we produced.

We know there may be good, and even compassionate, reasons for flying and driving. But do you consider how these actions might harm others? All are equal in Christ (Galatians 3:28), including those we harm by our actions.

Martin Luther King Jr on the good Samaritan

A true revolution of values will soon cause us to question the fairness and justice of many of our past and present policies. On the one hand we are called to play the good Samaritan on life's roadside; but that will be only an initial act. One day we must come to see that the whole Jericho road must be transformed so that men and women will not be constantly beaten and robbed as they make their journey on life's highway. True compassion is more than flinging a coin to a beggar; it is not haphazard and superficial. It comes to see that an edifice which produces beggars needs restructuring. A true revolution of values will soon look uneasily on the glaring contrast of poverty and wealth… and say, 'This is not just.'[2]

How compassionate are our journeys?

The parable includes journeys. The robbed man was presumably walking. The upper-class priest might have been riding an animal. The less well-off Levite might have been walking. The Samaritan travelled with a donkey.

Journeys made by walking or by riding animals have a small **carbon footprint**. What about other types of transport, like the car and the plane in our retold parable?

Transport accounts for about a third of **carbon emissions** in both the UK and the USA.[3]

You can calculate the **carbon footprint** for different forms of transport by using the free **carbon footprint** calculator at **climatestewards.org/offset**.

Flying

We recommend trying to reduce the number of flights you take, and avoiding flying when possible. Lots of people are trying that – the Swedish call it 'flygskam'. Most people in the world cannot afford to fly anyway.

How would you respond to someone who says, 'Flying is so cheap. Everyone else is having holidays in the sun. Why shouldn't we?'

___

___

___

___

A possible response:

Flying has very high **carbon emissions**. Virtual tours show more of the world at its best than we could ever hope to visit. Could you go on holiday by train instead? Could you at least try to reduce how frequently you fly? Perhaps you could petition the government for greater subsidies on train/bus travel.

What Jamie did

I signed the pledge at **flightfree.co.uk** to be flight free in 2020. And I have kept to it.

Train, bus, boat and bike

Travelling by train or bus is generally much less damaging to the **environment** than flying or using a diesel or petrol car, unless the train or bus is almost empty. You might see more on the way too. The website **seat61.com** has useful information about long-distance train travel. Boats can be a good option for crossing seas, though cruises have high **carbon footprints**.

Pedal bikes are a great option for short journeys, with a very low **carbon footprint**.

Motorbikes generally use less fuel than cars and require fewer raw materials to make. But although they emit less **carbon dioxide**, they emit higher levels of hydrocarbons, carbon monoxide and nitrogen oxides.[4] Electric motorbikes and electric bikes (e-bikes) are a better option for the **environment**.[5]

Cars

The website **nextgreencar.com** is good for checking the environmental impact of different cars. Diesel engines emit less **carbon dioxide** and are more fuel-efficient than petrol, but the soot particulates and noxious gases they emit contribute more to climate change, especially when driven at low speeds. Hybrid cars use an electric battery that is charged when they are run on petrol, and have better environmental ratings. Electric cars are better than all three but not perfect. Their **carbon footprint** depends partly on the **energy mix** of the power source where they are made and driven. Mining the rare metals used to make them often harms the **environment**. But over the lifetime of a car, electric cars emit around three times less **carbon dioxide** than the average petrol or diesel car.[6] The **carbon footprint** is much lower when the **energy mix** is more **sustainable**, and will drop further as batteries become easier to recycle and second-hand electric vehicles become more available.

Giving up cars altogether is even better. Giving up driving is less difficult if you live in a city with good public transport. Self-driving cars could end up replacing car ownership sooner than you think.[7]

Did you know?

If you are thinking of using an electric vehicle in the future, you can compare different types (including the cost) at ev-database.uk. As well as being greener, electric vehicles are much cheaper to run than other vehicles, with low fuel costs and simpler maintenance. Owners do not currently have to pay tax on fully electric vehicles in the UK (if they cost below £40,000), and there are far fewer parts to maintain.

What young people are saying

Kristina Stead, 18, and her brother Daniel, 14, recommend electric cars. On the left is an MG ZS with a sunroof. They describe this as one of the best-value electric cars, with lots of space. The battery has a 'real world' range of 140 miles.

Simon Stead

The car on the right is a Renault Zoe, with a 'real world' range of 195 miles. Both cars have rapid charging, which is useful during long journeys.

Changing our travel plans

The Samaritan had the compassion to change his travel plan to help an injured man. Our flights and diesel/petrol car journeys contribute to warming the atmosphere. The warmer the earth gets, the more people will suffer and die early. Do we have the compassion to make changes, to help those who are most at risk of the devastating effects of climate change?

We can ask the following questions about our journeys:

1 Is this journey necessary? Could you choose a holiday nearby rather than travelling a long distance? Could you join a meeting or conference by

internet instead of travelling? Could you walk to local shops instead of driving further away? Are you making this trip because you covet other people's trips?

2 Could you use a form of transport which has a lower **carbon footprint**? For example, walking, cycling, tram or bus instead of a short car journey; train or bus instead of a long car journey; train, bus or ferry instead of plane?

Did you know?

There is a range of bikes that are adapted for people with different disabilities. Even people in wheelchairs can join in family bike rides.

Sharon Pomeroy

3 If travelling by car, could you share the journey with others instead of travelling in separate cars? See **liftshare.com**. Is an electric car an option? If you drive, where possible drive in a way that will minimise your emissions (e.g. keep speed below 60 mph and slow down naturally with hills rather than braking fast). How else can you reduce the **carbon footprint** of car use?

4 If you have to fly, could you take one direct flight instead of changing? Direct flights are better for the **environment**, as a lot of the **carbon emissions** occur during take-off and landing.

If, after considering all of this, the **carbon footprint** is still high, you could try **carbon offsetting**. This means doing something to compensate for the **carbon emissions** that are produced. This could involve planting trees or donating to a project providing wind farms, efficient cooking stoves or hydroelectric dams, protecting forests or investing in **sustainable** technology development.[8]

It is better to avoid producing the carbon in the first place than to produce it and offset it. Why? The carbon price may be unrealistically low, and offsetting schemes may not be as effective as they claim. **Carbon offsetting** can easily become **greenwashing**, giving us licence to carry on harming the **environment** while deluding ourselves that we are not. However, if **carbon emissions** are inevitable, then it is better to offset than to ignore them. You can offset emissions at **climatestewards.org/offset**.

Hints from the Hawkers

Debbie and David hitch-hiked to Scotland (and back) on their honeymoon, getting lifts from a coach of girls on a hen party, lorry drivers and a sewage diver. As these people were all travelling anyway, there was little additional carbon footprint.

How could you use your travel choices to show love for your neighbours in nations devastated by climate change?

Climate refugees

Did you know?

Climate refugees are people who have had to leave their home region because of natural disasters or changes to the climate. Since 2008, an average of one person every second has been forcibly displaced from their home region by floods, storms, droughts or earthquakes. This number has been rising in recent years and is expected to continue to rise. Many walk for long distances to find safety.[9]

Refugees

They have no need of our help
So do not tell me
These haggard faces could belong to you or me
Should life have dealt a different hand
We need to see them for who they really are
Chancers and scroungers
Layabouts and loungers
With bombs up their sleeves
Cut-throats and thieves
They are not
Welcome here
We should make them
Go back to where they came from
They cannot
Share our food
Share our homes
Share our countries
Instead let us
Build a wall to keep them out
It is not okay to say
These are people just like us
A place should only belong to those who are born there
Do not be so stupid to think that
The world can be looked at another way

Brian Bilston[10]

Now read the poem from bottom to top. Which way of reading the poem do you agree with?

List people in the Bible who had to leave their homes, like refugees. If you need help, see Genesis 3:23; 7:7; 12:4–5; 16:6; 37:28; Exodus 2:15; 12:51; Ruth 1:1; 2 Chronicles 36:20; Matthew 2:13; Acts 8:1.

Who is a modern-day equivalent of the innkeeper?

Even the good Samaritan did not take the injured man home with him. He did not try to do everything by himself. After doing what he could, he paid someone else to do more than he could provide alone.

We can join forces with other people who will be more effective than us. Who might be an 'innkeeper' you could give money to, to do more about **climate justice** than you could do by yourself?

There are many groups acting for **climate justice** (see Appendix C). Some run projects which offset carbon. Others persuade governments to introduce policies to reduce **carbon emissions**. There are also many wonderful charities that help those who are suffering around the world because of climate change and other injustice. We have highlighted a few in this book.

Whatever you do, take effective action, and don't be like the priest and Levite who ignored the injured man.

God of compassion,
give us compassion for others when we make choices about our travel.
Help all who are refugees, including those who have left their homes because of climate change. Show us how we can help them. Amen

Loving our neighbours worldwide: **HoverAid in Madagascar**

uk.hoveraid.org

HoverAid

One Christian charity which helps people who are often ignored is HoverAid. HoverAid brings aid (including food, seeds, water pumps and building kits) by hovercraft to remote places. Those who receive it are grateful that, like the injured man in the parable, they no longer have to suffer alone. HoverAid also transports doctors and nurses to isolated communities with no access to healthcare, typically treating 500 patients a week.

A hovercraft is a vehicle that glides by hovering upon an air cushion. It can travel over water, land, mud, ice or sandbanks and so can go places inaccessible by other means of transport. It is particularly useful in flood relief work. Floods, hurricanes and cyclones destroy homes and communities.

A small hovercraft is able to reach these communities, creates relatively little pollution and has low **carbon emissions.**[11]

HoverAid works in the island of Madagascar. Madagascar has low **carbon emissions**, but suffers the effects of climate change (see Appendix B). As one of the areas most exposed to extreme weather events, Madagascar experiences on average three major natural disasters per year.[12]

Young people around the world: Climate refugees displaced by floods in Madagascar

After a week of rain, two children were woken in the night by their parents and taken on to the roof of their home. The river was rising rapidly around the house. At 8.00 am, with the river swirling around the rafters of the house close to where they were huddled, their prayers were answered. Paddling hard against the current, a large canoe approached, and the family were allowed to board. They collected 16 members of their extended family, and their seven neighbours.

Hoveraid

Heavily laden with people and possessions, the canoe began a treacherous journey in water chocked with debris and the corpses of livestock. It was scary. They reached a place of safety, where HoverAid came alongside them, providing food and seeds so that they could begin again.

Jamie's tips

The message here is simple – do something about it! There is no Planet B, so we need to do something to stop the destruction caused by climate change before the consequences become more devastating. Keep trying. Here are some ideas to start with.

1. Pray for **climate refugees** and other people suffering from floods, **drought** or other effects of climate change.

2. Watch the video at youtube.com/watch?v=zr8U28Wox5A about the **carbon footprint** of different forms of transport.

3. Find out more about how climate change is affecting people around the world. Don't be like the priest and the Levite, who ignored suffering. It is hard to love our neighbours if we don't know anything about them. Read the case studies in this book about young people affected by climate change. Think of them as your neighbours and your family.

4. Choose one of the charities in this book or a group that takes action on climate change (the modern-day innkeepers, Appendix C). Support them with your prayers, time or actions (3 John 5–6).

5. Walk, cycle or use public transport instead of a car for short journeys (unless you have an electric car). Walking and cycling are also good for your health.

6. If you travel by car, keep the windows closed when travelling above 40 mph, as driving with the window open increases drag and uses more fuel. Avoid using air conditioning.

7 If you drive, turn your engine off when you stop for more than two minutes (e.g. when waiting). If you're not the driver, politely ask the driver to switch the engine off.

Did you know?

In the UK, Rule 123 of The Highway Code looks at 'The driver and the environment', and states that drivers must not leave an engine running unnecessarily while the vehicle is stationary on a public road. This is known as 'idling', and it increases the amount of exhaust fumes in the air. These fumes contain harmful gases that contribute towards climate change and are linked to asthma and other lung diseases. The Royal College of Physicians estimates that 40,000 deaths a year in the UK are linked to air pollution, with engine idling contributing to this.

Electric cars that are switched on also engage in idling, and the electric battery is slowly drained. This is much less damaging than the idling of petrol or diesel cars, but could also be prevented by switching the engine off.[13]

8 Organise a sponsored walk, bike ride or other event to raise money to help HoverAid assist people in Madagascar.

9 Reflect on journeys you plan to take this year. Make changes to reduce the **carbon footprint** of these. Take trains and buses when you can, instead of flying or driving.

What Jamie did

Our school was already called an 'eco-school', but I believed we could do more. I wrote to my head teacher and asked him to make sure that school trips to Europe would use train, bus and ferry and not planes.

10 Encourage your friends, family members, church members, school or colleagues to take action to help the planet.

11 Write letters to your local council, Member of Parliament or other political leaders, asking them to introduce better policies to protect the planet. This could include requesting more charge points to encourage the use of electric vehicles or requesting more buses that run off electricity or bio-gas.

12 Make a commitment not to fly this year. Watch a 90-second video about this at youtube.com/watch?v=YfzlsUHLxxc.

Which of these do you already do?

____

____

____

Which will you attempt to do this month?

____

____

____

8 A time to plant

Bible passages

ECCLESIASTES 3:1–2

There is a time for everything,
and a season for every activity under the heavens:
a time to be born and a time to die,
a time to plant and a time to uproot.

DEUTERONOMY 20:19

When you lay siege to a city for a long time, fighting against it to capture it, do not destroy its trees by putting an axe to them, because you can eat their fruit. Do not cut them down. Are the trees people, that you should besiege them?

GENESIS 21:33

Abraham planted a tamarisk tree in Beersheba, and there he called on the name of the Lord, the Eternal God.

Plants and trees in the Bible

These Bible passages are short. But trees and other plants appear throughout the Bible. In the first two chapters of Genesis, we read that God created different kinds of plants and set Adam and Eve in a garden of trees. The final chapter of the Bible describes the tree of life, saying, 'The leaves of the tree are for the healing of the nations' (Revelation 22:2).

In Ecclesiastes 3:2, we read that there is 'a time to plant and a time to uproot'. We cannot harvest unless there has been a time of planting. Then, when a plant has come to the end of its life, it may be time to uproot it and begin the cycle of planting again.

Deuteronomy 20:19 instructed the Israelites not to cut down trees when they captured a city, because trees are useful, producing fruit.

What are trees useful for today?

____

____

____

____

Cherry tree, Penhurst Retreat Centre

Jay Ashworth

How do trees help the environment?

1. Trees absorb sunlight and use it to turn **carbon dioxide** and water vapour into food and oxygen, thus removing two **greenhouse gases** from the atmosphere and cooling the earth.
2. Tree roots prevent rain washing soil away, thus protecting the soil and preventing flooding.
3. Trees provide shade in the sunshine which helps to cool anything below them.
4. Trees absorb air pollution, purifying the air.
5. Trees help control regional rainfall as water evaporates from their leaves.[1]
6. Trees provide shelter and food for insects, birds, animals and plants.

7 Fruit trees provide food that, especially when eaten locally, is a low-carbon produce.

8 Trees can enhance well-being for people. Research shows that within minutes of being surrounded by trees, people experience a reduction in blood pressure and stress levels. Improvements in mood, sleep and energy levels are also reported.[2]

These are just some of the benefits of trees. You might be able to think of more.

Did you know?

Some trees absorb 22 kg of carbon dioxide each year. Trees can live for thousands of years. The different parts of a tree grow at different times throughout the year. Typically, most of the leaf growth happens in the spring. The trunk grows in the summer and roots grow more in the autumn and winter.[3]

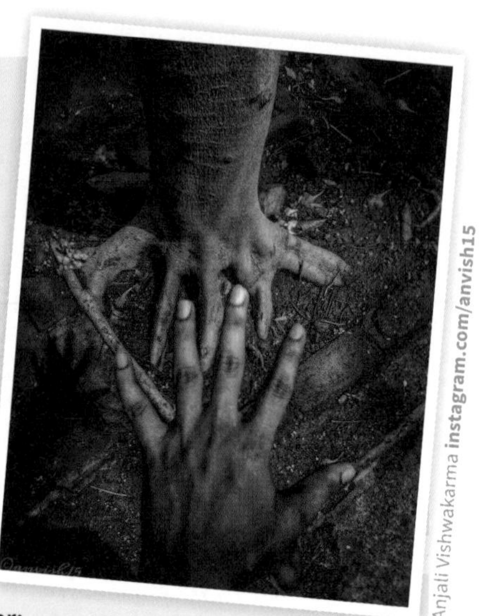

Anjali Vishwakarma **instagram.com/anvish15**

Words attributed to Martin Luther:
'Even if I knew that tomorrow the world would go to pieces, I would still plant my apple tree.'

Planting trees as altars of remembrance

Turning to our third Bible passage, in Genesis 21:33 we read that Abraham planted a tamarisk tree and called upon the name of the Lord. Tamarisk trees are evergreen and can grow very tall (up to 18 metres/60 feet). They provide shade and cooling. Abraham had just made a peace treaty with Abimelech, sorting out their disagreements. Abraham 'called upon the name of the Lord, the Eternal God', and planted a tree as a reminder that he was not acting on his own but depending on God.

Elsewhere in the Old Testament, there are several occasions where we read that an altar or a stone was set up to remind future generations of something God had done in that place (Genesis 28:18; 33:20; 35:7, 14, 19–20; Exodus 17:15; 24:4; Joshua 4:1–9, 20–22).

We can plant a tree to remind us of significant events, to thank God or to remind us of what God has taught us or done for us. We can also plant a tree in memory of someone who has died. Trees have long been viewed as appropriate memorials for the dead – for example, in 1 Samuel 31:13 we read that the bones of King Saul and his sons were buried under a tamarisk tree.

We can also plant trees to celebrate a birth, baptism, engagement, marriage, birthday or any other event. If you are struggling to think of a present for someone who seems to have everything they need, why not plant a tree for them (or send a donation to a charity who can plant one where it is needed)?[4]

Hints from the Hawkers

When we (Debbie and David) got engaged, our friend Roz gave us money to plant a tree to celebrate our engagement. Later, we drew up an 'alternative gift list', inviting our wedding guests to contribute to items for a trauma counselling centre for refugees, instead of giving us wedding presents. We included fruit trees on the list. With the donation from Roz plus donations in lieu of wedding presents, we were able to send money for fruit trees to be planted outside the counselling centre in Russia. Trees which reduce carbon emissions and provide fruit and clean air seemed a much better (and longer-lasting) wedding present than a toaster.

Where should we plant forests?

Planting forests outside the tropics (e.g. in Europe and North America) may warm the planet instead of cooling it down. That is because forests in these areas make the earth's surface darker so it absorbs more heat from the sun, especially evergreen trees in the winter which stop snow underneath reflecting sunlight. Tropical rainforests, in contrast, have lighter coloured leaves and grow so fast that they have a cooling effect.[5]

Rachel Haddock

Protecting existing trees

Planting trees is great, but they take years to mature and while they are young they don't absorb so much **carbon dioxide**. Sadly, some offsetting schemes have cut down trees to make space for new trees they were funded to plant. That is why we focus on projects like A Rocha, which preserve as well as plant, for instance in tropical rainforests.

The largest rainforest is the Amazon. But it is being destroyed, with a fifth of it already lost. Half of the world's rainforests have been destroyed in the past century. Trees are cut down to provide timber, fuel, paper and other products, as well as land for agriculture, cattle, mining and oil extraction.

When trees are cut down and rot, they return **carbon dioxide** back into the atmosphere. Deforestation is considered the second major driver of climate change (more than the entire global transport sector), responsible for 18–25% of global annual **carbon emissions**.[6] The **IPCC**'s Special Report on Climate Change and Land confirms that tropical forests play an important role in tackling climate change, so it is important to protect existing rainforests.

Did you know?

The Amazon rainforest is about half the size of Europe. Between 90 and 140 billion metric tons of carbon is stored in the Amazon forest.

If current trends continue, deforestation could double to 48 million hectares between 2010 and 2030, meaning that more than a quarter of the Amazon would be without trees.

Worldwide every year an area of forest the size of England and Wales is destroyed.[7]

Wetlands

Tyler Butler on Unsplash

It's not just forests that matter. Wetlands are areas where water often covers the soil (e.g. marshes and swamps), teeming with wildlife. Wetlands help protect us from flooding and clean our drinking water. About 90% of Europe's wetlands have been destroyed in the past 100 years.[8] Freshwater wetlands cover 1% of the earth's land surface but contain 40% of its **biodiversity**. Peat wetlands cover 3% of the surface and store a third of the world's carbon – more than all rainforests combined. Although wetlands also emit most of the world's **methane**, overall they help slow down climate change.[9]

Gardens

Changing how gardens are used can improve their environmental benefit. For ideas about how gardeners can help in the light of climate change, see **rhs.org.uk/advice/garden-health/wildlife/gardening-in-a-changing-climate**.

Leaving parts of a garden to grow wild increases the diversity in the garden. 'Rewilding' is becoming increasingly popular.[10]

Lawns have good drainage and store carbon, but mowing and maintaining a perfect lawn can cancel out their environmental benefits. It can help the **environment** if lawns are mown less often.[11]

Rainwater can drain through decking and paving, but will run off concrete and asphalt and contribute to flooding. Ideas for eco-friendly driveways can be found at **homeguides.sfgate.com/make-environmentally-friendly-driveway-77803.html**.

Allotments, plant pots and other ideas

If you don't have a garden, perhaps you could chat to someone who does (like a grandparent) about their garden or contact the council about the use of parks. Or maybe you can volunteer as a gardener for your church or community.

Alternatively, perhaps you could plant something in an allotment or in a hanging basket or plant pot indoors or outdoors. See tips 8 and 9 (below) for ideas.

Lord God, the true gardener,
We thank you for creating wonderful trees and plants. We are in awe at their beauty and all the ways they improve the environment. Help us to do what we can to protect forests and to appreciate and care for all that you have made. Amen

Loving our neighbours worldwide: A Rocha

arocha.org

Planting algarroba seedlings for reforestation

A Rocha is a Christian charity working in practical conservation projects around the world, including preserving wetlands and protecting and restoring tropical forests. For example, A Rocha's African Forests Programme aims to protect 50,000 hectares of **biodiversity**-rich tropical forests across Kenya, Uganda, Nigeria and Ghana. A Rocha also has forestry projects in Brazil and Peru.

In Peru, 29,000 hectares of the coastal dry forests are cleared every year, with species of trees becoming endangered and valuable **biodiversity** lost. Less than 5% of the dry forests remain, making this an international conservation priority. A Rocha Peru works with community members to plant native species of trees. A Rocha Peru also arranges patrols to locate and stop people who are illegally cutting down trees.

Workshops are held to train community members in dry forest restoration and conservation, including the production of traditional crafts using **sustainable** dry forest resources. Community members are also trained to produce a profitable syrup made from the algarroba (carob) tree and to use **ecological** ovens, which reduce the use of firewood and charcoal.

Young people around the world: Planting, pollution and plastic in Peru

Oliver is from Lima, Peru. Although he is only 10 years old, his life has already been affected by environmental issues. He explains:

'In Lima, there are too many cars that make lots of air pollution and this can make people sick. We need to plant more trees and stop cutting down the rainforest, because the trees help us by cleaning the air.

'Also, it is getting hotter. Some days in winter it is hot, and in the beach season it can be really hot and my skin gets burnt.

'Sometimes, in the summer when I go to the beach there is a lot of rubbish in the sea. One time, I was swimming, and there were plastic bags and food. I don't really like swimming with rubbish. I would like the seas and oceans to be cleaner. We need to stop throwing rubbish in the rivers and seas. We should all help to clean up.'

A Rocha, Peru

Oliver and his mum Filipa taking part in a beach clean-up

Oliver does what he can to protect the **environment**: 'We plant trees and have a garden. At school, we have an environmental council. I did a beach clean-up south of Lima with my mum. We collected lots of garbage like plastic cups, cigarette butts and plastic bags. We found a dead bird on the beach. I think it was a pelican. It was sad that the animal had died.'

A Rocha Peru organised the beach clean-up. 35 kg of rubbish was collected in one day.

Jamie's tips

1. Find some more Bible verses about trees. Start with 1 Chronicles 16:33; Daniel 4:10–12; Ezekiel 47:12; Habakkuk 3:17–18.

2. Go for a walk or bike ride and pay attention to the trees and plants that you see.

3. Trees are cut down to make paper. Recycle paper, to reduce the number of trees needing to be cut down. Try to use recycled paper, use both sides and only print when it is necessary.

4. Save envelopes from birthday and Christmas cards. Reuse them when you need to post something – you can score out your name or put a sticker over it.

5. Reuse wrapping paper for presents or be original and use newspaper or cloth. Let people know that you like reused wrapping paper on your gifts.

Did you know?

- **In the UK, 227,000 miles of wrapping paper is thrown away each year (enough to go round the earth nine times)!**
- **Approximately 50,000 trees are cut down to make the 8,259 tonnes of wrapping paper we use at Christmas.**[12]

6. Read about what different amounts of money could pay for if you donate to A Rocha Peru. See **globalgiving.org/projects/help-communities-to-conserve-dry-forests/photos/#menu**.

7. When paper or toilet paper is next bought for your household, choose some which is labelled as better for the **environment**. Look for the Forest Stewardship Council (FSC) mark. Or order 100% recycled paper from **uk.whogivesacrap.org**.

Did you know?

The Forest Stewardship Council (FSC) mark means that the paper or wood comes from responsibly managed forests. This applies to toilet rolls, paper, books and wooden furniture.

Products that bear the Rainforest Alliance Certified seal are sourced from forests that:

- **protect endangered species and forest areas of high conservation value**
- **set aside a portion of land as forest reserve**
- **provide workers with decent wages**
- **follow FSC guidelines that determine where forest products are harvested**
- **respect the rights of local communities and indigenous people.**

See products at rainforest-alliance.org/find-certified

8 Plant vegetables or soft fruits (like berries). If you don't have a garden or allotment, try growing them in pots. Potatoes can be grown in pots indoors – plant some potato peelings in soil. Cress can be grown in a dish lined with wet cotton wool or kitchen towel.

Hints from the Hawkers

We rinsed out a tin of tomatoes and threw the water in a plant pot to water the plant. Some months later, a tomato plant grew, presumably from seeds which had been in the tin.

9 Bees are an essential part of the **environment**. They pollinate many food crops and plants, but the number of bees has been falling recently (partly due to climate change and pesticide use). Boost their survival rates by planting

Verna Langrell

flowers like sunflowers, crocuses or lavender, or flowering herbs or beans (they are a bit fussy and don't like French beans, though). If you don't have a garden or allotment, put a hanging basket or a pot plant outside, or ask the council to plant flowers that bees like.

10 Give someone a plant, seeds or gardening equipment for their birthday or Christmas.

11 For your next birthday or Christmas, ask someone to send a donation to A Rocha to plant trees, instead of giving you a present.[13]

12 Ask your church, school, workplace, village, town or city to take A Rocha as their charity to fundraise for. Would they twin with an A Rocha project – not travelling there, but connecting remotely and raising funds?

Which of these do you already do?

Which will you attempt to do this month?

9 What shall we eat?

Bible passage

ROMANS 14:12–22

So then, each of us will give an account of ourselves to God.

Therefore let us stop passing judgement on one another. Instead, make up your mind not to put any stumbling-block or obstacle in the way of a brother or sister. I am convinced, being fully persuaded in the Lord Jesus, that nothing is unclean in itself. But if anyone regards something as unclean, then for that person it is unclean. If your brother or sister is distressed because of what you eat, you are no longer acting in love. Do not by your eating destroy someone for whom Christ died. Therefore do not let what you know is good be spoken of as evil. For the kingdom of God is not a matter of eating and drinking, but of righteousness, peace and joy in the Holy Spirit, because anyone who serves Christ in this way is pleasing to God and receives human approval.

Let us therefore make every effort to do what leads to peace and to mutual edification. Do not destroy the work of God for the sake of food. All food is clean, but it is wrong for a person to eat anything that causes someone else to stumble. It is better not to eat meat or drink wine or to do anything else that will cause your brother or sister to fall.

So whatever you believe about these things keep between yourself and God. Blessed is the one who does not condemn himself by what he approves.

Food for thought

What do you think people should eat? Are there any foods you think that people should avoid? Why?

____

____

____

____

Food rules in the Bible

Food rules changed throughout the Bible. God told the first humans that they could eat plants and fruit (Genesis 1:29). After the flood, God said people could eat animals, but not meat with blood in (Genesis 9:3–4).

Kosher food rules (e.g. Exodus 34:26; Leviticus 11) made foods like the following 'unclean' for Jews to eat: rabbits, pigs, shellfish, owls, most winged insects, lizards and goat cooked in its mother's milk.

In the New Testament, things changed again. There's no sign that Jesus broke Jewish food laws or avoided eating fish or meat (Matthew 14:16–21; Luke 22:7–8). But he said, 'What goes into someone's mouth does not defile them, but what comes out of their mouth, that is what defiles them' (Matthew 15:11). In other words, it's what we say (and do) that matters, not what we eat.

In the book of Acts, Peter had a vision of animals considered 'unclean' and heard a voice saying, 'Do not call anything impure that God has made clean' (Acts 10:15). In Acts 15:29, the apostles and elders agreed simple food rules for Gentile Christians: 'Abstain from food sacrificed to idols, from blood, [and] from the meat of strangled animals.' Paul wrote that the part about idols applied only if they knew that the food was sacrificed to idols (1 Corinthians 10:27–28).

- Do the changes tracked in the Bible show we should stick with one set of rules? If so, which one? Kosher rules? The early church's rules? No rules at all? Or does the Bible show that food guidelines change as conditions change?

Food choice is controversial. Some may think our suggestions don't go far enough. Others might think they go too far. Let's remember the words of the apostle Paul: 'Each of us will give an account of ourselves to God. Therefore let us stop passing judgement on one another' (Romans 14:12–13).

Paul wrote that he was convinced that no food was 'unclean' or banned any longer. He also acknowledged that some people thought otherwise. He wrote that we should not judge people who disagree (v. 13), nor upset anyone by what we eat (v. 15).

Quoting those who said they had the right to do anything, Paul responded, 'Not everything is beneficial' (1 Corinthians 6:12). Having the right to eat certain food does not make it a good idea. What we eat should be between ourselves and God.

Jesus said, 'Do not worry, saying, "What shall we eat?"... But seek first [God's] kingdom and his righteousness' (Matthew 6:31, 33). We each need to decide what we believe to be right and not judge others who hold a different view.

How can we seek righteousness in our choices of food? Here are some principles to consider:

- Don't be greedy. Proverbs 23:20 urges, 'Do not join those who drink too much wine or gorge themselves on meat.'
- Eat healthily, ensuring that you eat enough of different nutrient groups (including sufficient protein, carbohydrate and fat). Remember that our bodies are temples of the Holy Spirit and we should honour God with our bodies (1 Corinthians 6:19–20), neither under- nor overeating.
- Consider the needs of others. Let's not cause others distress because of what we eat (Romans 14:15).
- Show care for God's creation (see chapter 3).
- Share with others, rather than taking more than our fair share of resources (Proverbs 22:9; Luke 3:10–11; Hebrews 13:16).
- Make loving choices that will not cause harm to others (Romans 13:10).

These principles bring us back to the question of climate change. If our choices lead to high **greenhouse gas emissions**, we are contributing to the suffering of others.

How does what we eat relate to climate change?

Food production is responsible for a quarter of all **greenhouse gas emissions.**[1] Meat and dairy alone have a huge impact, with livestock accounting for about 14.5% of emissions.[2] That is similar to the emissions from all the world's cars, trucks, planes and ships combined. Cutting down forests to graze cows and growing extra grain to feed them vastly increases **greenhouse gas emissions**. Meat products have greater **carbon footprints** per calorie than grains or vegetables, especially because animals like cattle, sheep and goats produce a lot of **methane** gas, which ends up in the atmosphere.

Did you know?

Every time a cow burps, a little puff of methane is released into the atmosphere. Manure (cow poo) produces even more methane. Methane is a powerful but short-lived greenhouse gas. Measured over 20 years, it warms the earth about 70 times more than carbon dioxide, but it lasts only around 12 years before turning into other gases. Human activities – like the growing demand for beef and dairy farming, and also rice and landfill – have added methane to the atmosphere, so there is now more than 2.5 times the amount there was in 1750.[3]

Should we all give up eating animals and become vegetarian or vegan? Not necessarily. For instance, in some communities the farmland is no good for growing crops but benefits from grazing animals, as a **sustainable** food source.[4] Substitutes like almond milk take a huge amount of water to produce and may cause the death of many bees used in fertilisation.[5] Rainforests are often destroyed to grow soya.[6] Rice paddy fields produce **methane**. Cultured meat grown in laboratories (if powered by **fossil fuels**) produces **carbon dioxide**, which lasts for thousands of years (unlike **methane**, see box).[7] Going vegan for two meals a day and eating an unrestricted diet, including meat, for the third would cut **greenhouse gas emissions** more than three times as much as giving up red meat entirely while continuing to eat dairy, but only half as much as a completely vegan diet.[8] EAT, the 'science-based global platform for food system transformation', has produced recommendations for how we need to change what we eat and how we produce it, if we are to feed the population sustainably by 2050.[9]

Methods of farming

We have not talked much about different farming methods. If you want to know more about genetically modified foods or how **industrial farming** can be made **sustainable**, check out **explorebiotech.com/gmos-feed-the-world** and **link.springer.com/article/10.1007/s13593-017-0445-7**

Deciding what to do is not simple. Here are some recommendations to consider prayerfully. Not everyone will follow all of these, and that is fine.

Recommendation 1: Try not to waste food

No matter how high a food's **carbon footprint** is, it is better for it to be eaten than thrown away. About 30% of the world's food goes to waste annually (see chapter 1). **Organic** matter in landfill sites produces 20% of all **methane** emissions, contributing to climate change.[10]

Hints from the Hawkers to reduce food waste

1 Plan your meals, shop from a list and shop regularly so that you do not overstock perishable food.

2 Store food in a fridge or freezer. For example, bread can be frozen and slices defrosted when needed so that it does not go off. Bananas, grapes and other soft fruits can also be frozen. Cheese lasts longer if stored in the fridge in an airtight tub or re-sealable pack.

3 Let everyone serve themselves. Encourage your family to take only as much as they will finish. Let them come back for a second helping if they want more.

4 Store leftovers for later (e.g. in the fridge or freezer) and remember to use them up or offer them to someone else.

5 If bananas are going soft, make a smoothie, or a banana cake for a neighbour.

6 If carrots or parsnips dry up, soak them in a tub of water so that they rehydrate.

7 Revive biscuits that have gone soft by putting them into a recently used, warm oven. Keep checking them as they won't need long. Don't try this with chocolate biscuits unless you want a sticky mess – just find some children if you need chocolate biscuits eaten!

8 If, after all of this, you still have food waste, then compost it (making fertiliser for the soil) instead of putting it in a landfill site.

For more hints, see **lovefoodhatewaste.com**.

How much food do you waste? What could you do to reduce food waste?

____

____

____

____

____

Recommendation 2: Eat less red meat

Meat eating is rising rapidly, especially in China and other developing countries, but mainly to catch up with rich countries. The UK supplies an average of 219g of meat per person every day[11] – nearly twice the recommended amount for a healthy diet. Red meat (especially beef and lamb) generally has far higher emissions than other protein sources. Try having smaller portions of meat; meat-free meals or days; eating red meat only occasionally; replacing beef with chicken; or going vegetarian or vegan.

Did you know?

Daniel fasted from meat for three weeks (Daniel 10:3). Some parts of the Christian church (especially Catholic and Orthodox) have had meat-free days for centuries, which makes a lot of sense when meat is scarce. Ethiopian Orthodox Christians fast from meat on Wednesdays and Fridays and many other days – a total of 250 days a year.

Mena Remedios

Roasted root vegetable madras with dhal and spinach served with brown basmati rice

Recommendation 3: Eat less cheese

Cheese also has a high **carbon footprint**, mostly because it is made of milk. About 10 pounds of milk are required to make one pound of cheese.

Christina Birkby

Parsnip soup, hummus, bread and salad

Next time you plan to eat cheese, think of whether you could try protein with a lower **carbon footprint** instead (e.g. beans on your baked potato; peanut butter, hummus or egg in your sandwich; a handful of nuts).

Many people are becoming vegans because of their concern about climate change. If that is not the choice for you, could you at least reduce your meat and dairy intake? What change could you make?

____

____

____

Recommendation 4: Choose foods not transported by air

Transport accounts for a small proportion of the **carbon footprint** of most food. So generally, buying locally is less important than the type of food and the way it's grown and produced. For instance, tomatoes grown in heated greenhouses in a cold climate may have a higher **carbon footprint** than tomatoes transported by sea from a warm climate. The exception is when food has travelled by air, as this greatly increases the **carbon footprint**. This is most likely to apply to fresh foods which perish quickly and have travelled a long distance. Fresh foods often have a label saying where they have come from, so you can see the distance travelled. Asparagus, green beans and berries that have been imported from a distant country have often travelled by air.[12]

Verna Langrell

Picking blackberries

If you have a garden, allotment or plant pots, try growing some of your own food, so that no transport will be needed. Or pick blackberries, apples or other fruit growing wild alongside public footpaths.

Recommendation 5: If you eat fish, choose a sustainable type

Overfishing and common fishing practices have led to a short supply of certain types of fish – often including popular fish, such as tuna and cod. Choose fish that are plentiful. The Marine Conservation Society regularly updates a list of which fish are most **sustainable** (depending on how and where they are caught – check the label or ask when you buy). See **mcsuk.org/goodfishguide**.

How sustainable is the fish you eat? Could you switch to more sustainable fish?

____

____

____

____

What about palm oil?

Palm oil is a cheap and common ingredient in many foods, such as biscuits and margarine, and cosmetics, such as shampoo, soap and toothpaste. In places like Indonesia and Malaysia, tropical rainforests have been destroyed to make space for palm tree plantations to produce palm oil. Cutting down existing trees to plant palm trees is bad for the soil, endangers species like orangutans and tigers and releases **carbon dioxide** into the air, speeding up climate change (see chapter 8). Fertilisers and pesticides, which are used to grow palm trees, wash into rivers and harm the health of wildlife and local people.

Working out what to do about palm oil is not easy.

- Palm oil is so common that it's hard to avoid, and it is not usually certified as **sustainable**.
- The UK shop Iceland has reformulated its products so that it no longer uses palm oil in any of its own-brand items.
- However, palm trees produce more oil than other crops on the same amount of land, so more forests might be cut down to produce sunflower, rapeseed or soybean oil.
- Palm oil certified as **sustainable** is produced by companies that have pledged not to clear any primary forest and must: have transparent supply chains; check how much carbon they are emitting; reduce the use of pesticides and fires; limit planting on peatlands; create wildlife zones; and treat their workers fairly (**rspo.org**). But research has shown that many of the certified '**sustainable**' palm plantations were sown in recently destroyed rainforests, which raises doubts about whether any palm oil can really be **sustainable**.[13]

Maybe the best we can do is put pressure on companies and governments to stop destroying forests to produce palm oil.

Asha Fitzpatrick, 12, and her sister Jia, 10, stopped eating Kellogg's cereals and began a petition asking Kellogg's to stop using unsustainable palm oil. Their petition gained more than 785,000 signatures. They did not initially receive the reply they wanted, so they persisted. Kellogg's has now stated that it has committed to sourcing palm oil only from suppliers and locations 'that uphold the company's commitment to protect forests and peat lands, as well as human and community rights'.

For more information about what you can do about this issue, see **palmoilscorecard.panda.org/file/WWF_Palm_Oil_Scorecard_2020.pdf**.

We have offered principles to consider. Stick to your convictions, but do not judge others who make different choices.

God our provider,
we thank you for our food. Please help us to enjoy it without being greedy or taking more than our fair share of the earth's resources. Guide us in our choices. Help us to be considerate to others and to show care for your world. Amen

Did you know?

Yemen is one of the poorest countries in the Middle East. It is one of the countries most vulnerable to climate change and food insecurity (see Appendix B). In 2014, a year before the Yemeni civil war began, the World Bank observed that the water table was sinking by six metres per year in the countryside, indicating a shortage of water. Experts predicted that the capital would use up its water supply by 2023.

Abdulhakim Aulaiah, a former official in the Environmental Protection Authority of Yemen, warned, 'Climate change has affected most aspects of life in Yemen… As a result of unusually high temperatures, malaria is spreading. Fluctuations in rainfall have affected crop yield across Yemen. The supply of fish in the seas around Yemen is decreasing, and several species have vanished.'

Fertile land has turned to desert. Drought leads to food shortages. Yemen has been hit by hurricanes, cyclones and sandstorms which damage crops and cause soil erosion.

Ban Ki-moon, the former secretary general of the United Nations, has said, 'Land degradation, caused or exacerbated by climate change, is not only a danger to livelihoods but also a threat to peace and stability.'

Even before the 2020 pandemic, UNICEF estimated that 537,000 Yemeni children under the age of 5 were at risk of severe acute malnutrition. Every second person in Yemen was struggling to find his or her next meal, according to the UN Office for the Co-ordination of Humanitarian Affairs (OCHA).[14]

Loving our neighbours worldwide: **AWM-Pioneers in Yemen**

awm-pioneers.org

AWM-Pioneers is a charity that uses donations to make an impact in Yemen through providing humanitarian aid (clean water and food), health work (including cholera prevention), education, training teachers and leaders, and helping victims of trauma.

AWM-Pioneers have contacts in Yemen who provide food parcels, drinking water and hygiene kits for those in desperate need. They have noticed a disturbing trend of older men buying girls (under 15 years old) as their wives, in return for providing food for their hungry families. AWM-Pioneers works with contacts who provide food to keep families together, so that children are not sold off.

AWM-Pioneers

Family receiving food aid in Yemen

Young people around the world: **Saved from being sold as a child bride because of hunger in Yemen**

Akira's (not her real name) father had died. Her mother had no income to provide food for the family. She was looking after Akira's brother, who has special needs. A man in his 50s offered to marry 14-year-old Akira and said that in return he would provide food for the rest of the family. Akira's mother says, 'I almost sold my daughter because of my hunger.' Thankfully, a local contact of AWM-Pioneers was able to provide food for the family, so Akira did not have to be sold. This kept the family together.

Jamie's tips

We should do what we can, not what we can't, so choose some actions which are manageable for you.

1. Thank God for your food before every meal this week. Not everyone has as much food as we do.

2. Climate change contributes to world hunger. Pray for people in Yemen and other countries where there are food shortages (see **prayercast.com/yemen.html**).

Photo and cooking by Iftekhar Ahmed

Mushroom salad with sesame sauce

3. Try some vegan food, such as nuts, beans, lentils, peas or tofu. See examples in the photos in this chapter.

4. Share some food with someone else, maybe crisps, chocolate or other snacks.

5. A calculator that works out the **carbon footprint** for different types of food is available at **bbc.co.uk/news/science-environment-46459714**. Use it to discover the **carbon footprint** of foods you enjoy. Work out whether you want to eat more or less of any foods based on this.

6 Check the food in your fridge or cupboard, and see if there is anything which needs to be used soon. Try to ensure that it gets used (or frozen or given away) before it goes off.

7 If you eat fish and chips (or any other fish meal), research to see if there's more **sustainable** fish than the type you currently choose. Consider choosing that next time.

8 Try to do food shopping without using a car. Is there a food shop you could walk, cycle or take a bus or tram to?

9 Eat less red meat. When you have a choice (e.g. at school or in a café, canteen or restaurant), choose a meal without red meat.

10 Become vegan or vegetarian for a day, week or month.

11 Don't waste any food this month. Only take what you will eat.

Dora Lit

Choi Sum (Chinese vegetable dish)

12 With your youth group or friends, invite people to a meal, and ask them to donate to the Yemen Appeal at awm-pioneers.org/giving. Try to follow the recommendations in this chapter, and make sure that leftovers are given away, used or frozen. See Arabian recipes and fundraising ideas at awm-pioneers.org/fundraise/fundraising-resources/arab-recipes. If a meal is too much to arrange, sell cakes instead.

What Jamie did

With my youth group, we provided a simple meal for people at church. With help from our youth leaders, we cooked jacket potatoes with beans or other fillings. There were only about four of us in the youth group, but we managed to raise £500 for charity in one lunchtime.

Which of these do you already do?

____

____

____

Which will you attempt to do this month?

____

____

____

10 Mustard seeds: small actions (and young people) matter

Bible passages

JOHN 6:5–13

When Jesus looked up and saw a great crowd coming towards him, he said to Philip, 'Where shall we buy bread for these people to eat?' He asked this only to test him, for he already had in mind what he was going to do.

Philip answered him, 'It would take more than half a year's wages to buy enough bread for each one to have a bite!'

Another of his disciples, Andrew, Simon Peter's brother, spoke up, 'Here is a boy with five small barley loaves and two small fish, but how far will they go among so many?'

Jesus said, 'Make the people sit down.' There was plenty of grass in that place, and they sat down (about five thousand men were there). Jesus then took the loaves, gave thanks, and distributed to those who were seated as much as they wanted. He did the same with the fish.

When they had all had enough to eat, he said to his disciples, 'Gather the pieces that are left over. Let nothing be wasted.' So they gathered them and filled twelve baskets with the pieces of the five barley loaves left over by those who had eaten.

MARK 12:41–44

Jesus sat down opposite the place where the offerings were put and watched the crowd putting their money into the temple treasury. Many rich people threw in large amounts. But a poor widow came and put in two very small copper coins, worth only a few pence.

Calling his disciples to him, Jesus said, 'Truly I tell you, this poor widow has put more into the treasury than all the others. They all gave out of their wealth; but she, out of her poverty, put in everything – all she had to live on.'

1 TIMOTHY 4:12

Don't let anyone look down on you because you are young, but set an example for the believers in speech, in conduct, in love, in faith and in purity.

Small actions matter

Can you think of any times in the Bible when something little was multiplied or treated as significant, or when seemingly small actions made a big difference, for good or for ill?

In chapter 1 we considered what the feeding of the 5,000 has to say about avoiding waste while being generous. In this chapter, we will consider another aspect of this miracle: that we can use the little we have for God, and God can multiply it.

The world is so big. What we do can seem insignificant. But God delights in using the small and insignificant.

The feeding of the 5,000 men (plus women and children) is the only miracle performed by Jesus that is recorded in all four gospels (Matthew 14:13–21; Mark 6:30–44; Luke 9:12–17; John 6:1–14). All four gospel writers must have thought it significant. A child was willing to share his packed lunch. Perhaps no adult thought it worth offering their little contribution when the needs were so great. Philip started trying to calculate the cost and thought, 'We can't do it.' But a child was willing to do what he could and offered to share his meal. Jesus took this and multiplied it, so that it fed the crowd and there were twelve baskets left over.

When you think about the climate crisis, do you tend to think it's too big a problem (like Philip) or that we should do what we can (like the boy)?

____

____

____

This miracle reminds us of Elisha, who surprised his servant by feeding 100 men with just 20 loaves of bread. As they were intended just for Elisha and carried by one man, they were probably small loaves, perhaps more like bagels. And yet 100 men were fed, and there was food left over (2 Kings 4:42–44). God can multiply small into large, as we see also in the case of the oil and flour that did not run out (1 Kings 17:7–16; see also 2 Kings 4:1–7).

Instead of looking at what we don't have or can't do, we can offer the little we have to God and ask God to multiply it. In the second passage, a poor widow put a tiny coin into the offering. It was all she had to live on. She could have held it back because she needed it herself or because it seemed too little to make any difference and it would be embarrassing if anyone saw what she was giving. Instead, she gave what she could. Jesus commended her, saying that this was worth more than the offerings of those who only gave what they could spare.

'Do not despise these small beginnings, for the Lord rejoices to see the work begin' (Zechariah 4:10, NLT). God welcomes small beginnings. If we only have one talent, God does not want us to hide it in the ground, but rather to use it (Matthew 25:14–30). Jesus said that faith as tiny as a mustard seed can move mountains (Matthew 17:20). God told Gideon that his army of 32,000 men was too large to fight 135,000 Midianites, so he chose a small handful (300) of the soldiers and they defeated the Midianites (Judges 7—8). God takes small things and weak people and multiplies their potential (1 Corinthians 1:27–28). That way we know that it is God at work, not us.

Young people matter

What do you think the Bible says about children and young people?

____

____

____

____

Paul wrote to the young man Timothy, 'Don't let anyone look down on you because you are young, but set an example' (1 Timothy 4:12). In God's eyes, young people matter. Young people, not just older people, will be filled with God's Spirit (Joel 2:28).

The disciples thought children were unimportant and did not want them to bother Jesus. Jesus rebuked the disciples and welcomed the children (Mark 10:13–16). Elsewhere, Jesus said that we should all 'become like little children' (Matthew 18:3). God is pleased with the praise of children (Psalm 8:2; Matthew 21:15–16).

Young people used powerfully by God include the child Samuel; Joseph son of Jacob, who was 17 when he had prophetic dreams (Genesis 37:2–9); the prophet Jeremiah (Jeremiah 1:6–7); Naaman's wife's slave girl (2 Kings 5:2–3); Esther;[1] the boy who shared his lunch; and Mary, who is thought to have been a teenager when she gave birth to Jesus.

The first king of Israel was Saul, who was from the smallest tribe in the smallest clan (1 Samuel 9:21). God chose David, the youngest son of Jesse, to defeat huge Goliath and later to become the next king (1 Samuel 16—17).

In a childlike way, David turned down the offer of heavy armour and chose a sling as his weapon. An adult would not have thought of that approach, but it turned out to be exactly what was needed. With a sling and five small stones, David was able to attack while out of reach of Goliath's weapons. He hit the only unprotected part of Goliath's body, his forehead, knocking Goliath to the ground. What stands out most is David's childlike lack of fear and his trust that God was with him: 'The Lord who rescued me from the paw of the lion and the paw of the bear will rescue me from the hand of this Philistine' (1 Samuel 17:37).

We might feel like we are tackling a giant when we try to persuade multinational companies, banks and governments to change. That is a good time to remind ourselves who won when little David fought Goliath.

Think climate!

David Hawker

When Debbie was young, she noticed road signs that said, 'Think bike!' She wondered why she was supposed to think about a bike. Was it suggesting, 'Think about travelling by bike instead of taking the car'? That would have been a good environmental message, but it is not what was intended. These signs are used to remind drivers to look out for bikes on the road and to pass them carefully.

We would like to see 'Think climate' signs and posters displayed as well, to remind us to think about climate change as we make decisions throughout the day.

One way to train ourselves to *think climate* is to keep doing small actions for the climate. These mustard seeds could grow into something bigger with us doing more and perhaps even inspiring and influencing other people. School climate strikes began small before becoming a mass movement.

What small, positive climate actions do (or can) you take to keep climate justice in your mind?

____

____

____

____

> **Greta Thunberg said:**
> **'Every single person counts.**
> **Just like every single emission counts.**
> **Every single kilo.**
> **Everything counts.'[2]**

Small steps to care for creation

Here are some ideas. We hope you can think of others. Choose whatever is relevant for you.

1 Put only as much water as you need in the kettle, so that you don't waste energy boiling water you won't use. You could boil enough to fill a vacuum flask and then use the flask throughout the day instead of boiling the kettle again.

2 Hoover less and iron less, to save electricity (this is one of our favourite ideas!).

3 Avoid sending unnecessary emails or copying people into emails if you don't need to. One email has a tiny **carbon footprint**, but that can multiply.

Did you know?

Every email we send uses electricity to display it, and the network connection uses electricity while the email is being transferred. As the email travels across the internet, each server will use some electricity to temporarily store it, before passing it on.[3]

Hints from the Hawkers

We used to send a Christmas email to about 1,000 people, to save carbon by not sending cards. When we realised that sending a large attachment has a carbon footprint, we decided instead to put our letter on our website and alert people to it, no longer emailing an attachment (to the delight of people who no longer felt obliged to read it, we suspect!).

4 Wash clothes less often, and wash them at 30 degrees rather than on a hotter wash.

5 Take opportunities to raise awareness about climate change. Pass the message on.

What Jamie did

Jamie takes opportunities to get people to think about climate change. As well as writing to the local paper about recycling and doing a talk on plastic waste for his youth group, he asked his school to include more about climate change on the curriculum. When his school announced a competition to write a book, he wrote and illustrated one about climate change. In English assignments, he has chosen to write stories about climate change. He has taken part in a 'die-in' (organised by Christian Aid and Extinction Rebellion, at Greenbelt festival) to highlight the death of the planet. And he has questioned an MEP about green policies.

6 Thinking of having a baby in the future? Plan to use reusable nappies.

Did you know?

About eight million disposable nappies are thrown away every day in the UK. They could take hundreds of years to decompose. Using cloth nappies for one baby saves an estimated 874 kg of nappy waste filling up landfill sites and reduces the carbon footprint by up to 40 per cent compared with using disposable nappies (if washed on a full load and hung out to dry, then passed on when your baby graduates from potty training school). You can also save a small fortune in nappy costs.[4]

What Jamie did

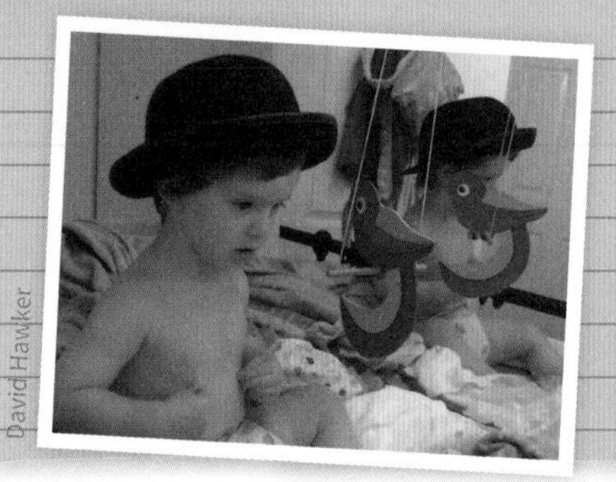

David Hawker

Although he was too young to make the choice himself, Jamie did not object to reusable nappies (received from Freecycle and re-gifted when Jamie finished with them).

7 Promote the use of reusable sanitary products. This is rarely talked about, but women can use reusable sanitary pads and tampons.[5] Debbie has been using these for over 30 years and finds that modern versions are discreet and stay in place well. Over the course of a lifetime, the average woman in a resource-rich country uses between 5,000 and 15,000 pads and tampons. The vast majority of these end up in landfill as plastic waste.[6] Throughout most of history

Simon Guillebaud

Providing reusable sanitary towels so girls don't give up school

and geography, women have reused cloth products. Although this is a choice for the girls, guys can help by donating to charities which provide reusable period products to girls in resource-poor countries.[7]

8 Consider how you wash your dishes. Washing dishes by hand does not necessarily use less water or have a lower **carbon footprint** than using a dishwasher, though it depends how you do it. If washing by hand, use a bowl of water instead of keeping the hot water running. If using a dishwasher, ensure it is full and choose an economy setting if available.[8]

9 If you use a face mask, choose a reusable one.

The book *How Bad Are Bananas?*[9] was published back in 2010 but is still very useful. It calculates the **carbon footprint** of items from text messages to a night in a hotel to being cremated to having a child, and about a hundred more besides. This information provides many ideas about ways to reduce our **carbon footprint**.

What young people are doing

When Molly Elliott was about 10, she entered some art works about climate change in a competition. She won £2,000 for her school, which helped purchase solar panels.

Years later, Molly attended a school climate strike. It took over two hours to get to her nearest strike by bus and train, so next time she liaised with the police and set up a strike in her own area. Four months later, 600 people joined the strike she organised. Molly was 15.

In the last year, Molly has completed a work experience placement with Friends of the Earth. She also keeps in touch with her local MP to keep climate change on their radar. She is 16.

Millie Elliott

Molly thinks it is a shame that it has to be children and young people who take a lead in these actions. She wants adults to take the initiative as well.

She hopes we can all find the right balance between making individual changes (e.g. food, transport, energy use and shopping habits) and working to change the system (e.g. by attending protests and using our vote with the future generation in mind). She states, 'As students and young people we want policies and legislation that force the government and companies to comply with the Paris targets of only 1.5 degrees of warming.'

Young people matter. Young people inspire us, and they are doing a lot. They give us hope that the next generation will take better care of the world. But that does *not* mean that the rest of us can ignore the problem. We all need to play our part.

The school strikes Molly organised are part of the UK Student Climate Network, which demands that:

1. the government declare a climate emergency and implement a Green New Deal;[10]
2. the education system is reformed to teach about the urgency, severity and scientific basis of the climate crisis;
3. the government communicate the severity of the **ecological** crisis and the necessity to act now;
4. everyone living in the UK over the age of 16 must have the right to vote in elections, conducted via proportional representation.[11]

Small steps should lead to bigger ones, which we discuss in the next chapter. Our small actions and our large actions make a difference, doing either harm or good. If we *all* do what we can, we will get somewhere. Drops in the ocean join together to form a wave which cannot be stopped.

Abba Father,
thank you that we are your little children. Help us to be faithful in the small things and the big things. Please multiply the little we offer and make it into something bigger. Thank you. Amen

Climate change can increase the risk of malaria and other diseases

An African proverb quoted by the Dalai Lama says, 'If you think you are too small to make a difference, you haven't spent the night with a mosquito.'

This little insect transmits parasites that cause malaria. Malaria caused over 400,000 deaths per year up to 2020 (mainly in children under 5). Many more people were seriously ill with malaria but recovered.

Rising temperatures and humidity increase the transmission of many diseases that are carried from one person to another via a third organism (like an insect), including malaria, dengue fever, encephalitis and yellow fever.

Professor Fernando in the *United Nations Chronicle* states, 'Climate change will increase the opportunities for malaria transmission in traditionally malarious areas, in areas the disease has been controlled, as well as in new areas which have been traditionally non-malarious... Increasing temperatures and global travel have the potential to reintroduce or increase transmission of malaria in tropical and temperate countries that have either eliminated or controlled transmission.'

Outbreaks of cholera and diarrhoeal disease can also be linked to climate change. Climate change also (directly or indirectly) increases the risk of epidemics and pandemics, such as Ebola and Covid-19.[12]

Egor Kamelev on Unsplash

Loving our neighbours worldwide: SIM in Pakistan

sim.co.uk

Serving In Mission (SIM) has workers in over 60 countries. The organisation's website states, 'The SIM family can use people of every ethnicity who have almost any skill imaginable!' This includes doctors, nurses and other health workers who work in hospitals and community health ministries across the world. Some of these treat cases of malaria and other diseases. Others provide health education, to try to reduce the risk of illness.

In Pakistan, SIM has a couple of medical teams who, at the invitation of local church partners, travel to villages to help families understand more about health and hygiene, wound care and common diseases.

SIM, Pakistan

Health education

Young people around the world: Malaria in Pakistan

A nurse with SIM in Pakistan received a call from a Christian friend. Her 2-year-old son, Elijah, had become sick with cerebral malaria. He had had a high fever for three days, but – perhaps because of high medical costs, perhaps because of busyness – his father had not taken him to hospital. By the time he was taken to hospital, he had lost consciousness and was in a critical condition.

The family and church gathered around Elijah, praying all night while the medics treated him. Looking at his medical notes, it was obvious that they had arrived just in time.

By God's grace, Elijah recovered. The nurse had the chance to sit with Elijah's dad. 'This child is a precious prince that you have been given by God,' she told him. 'These diseases are serious and need rapid treatment.' Elijah's dad had tears in his eyes. Many people do not know about the dangers of malaria and other treatable diseases. It was a joy to join Elijah's dad in praising God that God had spared Elijah.

Jamie's tips

These are written in no particular order as, unlike other chapters, I think they are all equally easy.

1. Turn off lights when you don't need them or are leaving the room (and use low-energy LED lightbulbs).
2. Shop locally, and try not to order online if you don't need to.
3. Put an Extinction Rebellion or 'Think Climate' poster in your window.
4. Pray to God, who hears our small prayers and 'is able to do immeasurably more than all we ask or imagine' (Ephesians 3:20).

5 Switch appliances and chargers off at the wall when not in use. You should do this anyway as a charger may cause a house fire if left on for many hours.[13] Don't leave appliances on standby overnight.

Did you know?

Any charger that is plugged in at the wall and not switched off at the socket will still use some electricity, even if it's not plugged into a device. A typical home could save around £30 a year by making sure that all chargers and appliances are unplugged, or switched off at the wall, when they're not being used. When any product is constantly plugged in, heat is produced and wasted. That can lead to the deterioration of the electronic components over time, so that the equipment does not last as long as it would if you turned it off, thus causing further waste.

6 Read the short book by Greta Thunberg entitled *No One Is Too Small to Make a Difference*. This contains her famous speeches from 2018–19.

7 Do small jobs, like gardening, to raise money for SIM or find another way to raise funds. SIM helps people who have diseases related to climate change.

8 Use less water by not flushing the toilet if it is just a wee. Only flush if it is a number two. 'If it's yellow, let it mellow. If it's brown, flush it down.' You might need to chat with other people in your household about this, but it would help save a lot of water. Flushing the toilet accounts for nearly a third of household water consumption in the UK, and parts of the UK (and many other countries) sometimes face water shortages.[14] Flushing less saves water and also saves the energy and resources which are needed to treat every flush, thus reducing **carbon emissions**. It helps us to keep *thinking climate* throughout the day, and keeps us aware of people who don't have easy access to water (and it might reduce the water bill).

Did you know?

An average one-person household in the UK used 149 litres of water per day in 2018.

In the UK, 3.68 million tonnes of carbon dioxide are produced each year by the supply and treatment of our water.

You can find more tips on reducing water waste at cat.org.uk/info-resources/free-information-service/water-and-sanitation/reducing-water-use[15]

9 Have a short shower rather than a bath if you can. A five-minute standard shower uses about a third of the water of a bath, according to the UK Environment Agency (although a five-minute power shower uses as much water as a bath). Use a water butt to collect water for gardens/allotments or collect water from your shower to water plants or even to flush the toilet.

10 If you stay in a hotel or guest house, tell staff that the towels and sheets do not need to be washed during your stay. Many hotels now have cards in the rooms promoting this option to help reduce **carbon emissions** caused by washing linen daily. Also, if shampoo, shower gel and water are provided in individual plastic bottles, avoid them to reduce plastic waste – take your own along with you.

11 Encourage teachers to put more climate change teaching into geography/science lessons.

12 Look for more ideas on websites, such as **insightsforchange.co.uk/lifestyle**, **wikihow.com/Reduce-Your-Energy-Consumption** and **threerivers.gov.uk/egcl-page/reduce-re-use-recycle-recover**. Or watch the YouTube videos listed in Appendix D.

Which of these do you already do?

Which will you attempt to do this month?

11 Moving mountains: large-scale action

Bible passage

MATTHEW 23:23–28

'Woe to you, teachers of the law and Pharisees, you hypocrites! You give a tenth of your spices – mint, dill and cumin. But you have neglected the more important matters of the law – justice, mercy and faithfulness. You should have practised the latter, without neglecting the former. You blind guides! You strain out a gnat but swallow a camel.

'Woe to you, teachers of the law and Pharisees, you hypocrites! You clean the outside of the cup and dish, but inside they are full of greed and self-indulgence. Blind Pharisee! First clean the inside of the cup and dish, and then the outside also will be clean.

'Woe to you, teachers of the law and Pharisees, you hypocrites! You are like whitewashed tombs, which look beautiful on the outside but on the inside are full of the bones of the dead and everything unclean. In the same way, on the outside you appear to people as righteous but on the inside you are full of hypocrisy and wickedness.'

What do you think this Bible passage is about?
Is it relevant today?

____

____

____

____

Missing the big picture

Lawyers and Pharisees are the bogeymen of the gospels. But they were trying hard to do their bit. Israel, occupied by the Romans, faced a national crisis. The Pharisees hoped that if people followed the rules, the Messiah would come and put everything right. The rules were hard to follow and apply, but the lawyers had worked out how to do it.

Jesus saw that they had missed the big picture. They gave away a tenth of their spices. Can you imagine sorting out a tenth of every little jar in the spice rack to give away? It would be a fiddly job. Virtue-signalling blinded them to the fact that they were doing more harm than good. Sidetracked by little rules, they 'neglected the more important matters of the law' (v. 23). Looking good on the outside, they were wicked inside (v. 28).

We know better than the lawyers and Pharisees. Or do we? Some might say:

> Woe to you, eco-warriors and environmentalists, you hypocrites! You recycle all your plastic, glass and paper, but you are still flying, driving and having large families. Recycling is great, but you need to change your life to make a bigger impact!
>
> Woe to you, eco-warriors and environmentalists, you hypocrites! You offset your **carbon footprints**, sign petitions to divest from **fossil fuels** and look like you are supporting the **environment**. But it is all hollow **greenwashing**. You are still voting for governments that keep a system going that makes climate change worse.
>
> Woe to you, eco-warriors and environmentalists, you hypocrites! You take small steps but go no further. You think you are doing well because you save water at home and switch off your phone chargers. You are blind to your actions that deprive people of water far away and to the rainforest destroyed to make your phone!

Are we hypocrites?

Saving the planet is tiring. There's always more to do, but we can't do everything. We might feel we've done enough. We might feel like giving up, especially if it turns out that our actions have made little difference or even made things worse. We might feel hypocritical for not doing more or for driving a car, taking a flight, buying clothes or how we eat.

Isn't it hypocritical to print a paper book on climate change? When Jamie was six, we read him an e-book, and his favourite part was at the end of each chapter: 'We care about e-books because we care about the **environment**. Read an e-book and save a tree. You can help save our planet.' Yet here we are providing a printed version as well as a PDF e-book (hoping that it will bring about positive changes that will outweigh the production of the book).

People are imperfect and inconsistent. But that does not make them hypocrites. Jesus called hypocrites those who saw the speck in someone else's eye while ignoring the log in their own eye (Matthew 7:5), and who were more concerned about looking good on the outside than being right on the inside. 'People look at the outward appearance, but the Lord looks at the heart' (1 Samuel 16:7).

Jesus told us to pay attention not to how we look to others, but to God's priorities – 'justice, mercy and faithfulness' (Matthew 23:23). In this chapter, we will consider each of these.

Justice: protecting the most vulnerable people

In chapter 3, we noted that the Bible keeps talking about justice. Two Hebrew words used are *mishpat* (often translated 'judgement') and *tsedek* (often translated 'righteousness'). We sometimes think of those words as being about imprisoning criminals and living blamelessly. But in the Bible, they're largely about making sure life is fair for everyone. For example: 'Learn to do right; seek justice [*mishpat*]. Defend the oppressed. Take up the cause of the fatherless; plead the case of the widow' (Isaiah 1:17).

Nuevas Esperanzas

Boy collecting water in Nicaragua

God kept asking his people to look after the powerless and vulnerable – orphans, widows and immigrants, who had no family to support them (Deuteronomy 10:18; James 1:27). Such people today are those who cannot work, have few resources and have no financial support from governments or family. They have few rights and no land they can farm. They need justice.

The Bible's emphasis on justice is echoed in international priorities. The United Nations has set 17 Sustainable Development Goals to achieve a more just future for everybody by 2030.[1] The **IPCC** has looked at how we can deal with climate change and improve **sustainable** development at the same time.

Mercy: what contributes most to climate change?

The usual Hebrew word for mercy is *hesed*, also translated 'loving kindness'. The small steps in the previous chapter are a good start, but if we want to be merciful to people suffering from climate injustice, we need to prioritise actions that will make the most difference to them.

What do you think contributes most to climate change?

____

____

____

____

____

____

____

The biggest contributors to **greenhouse gas emissions** are electricity, heating and transport powered by **fossil fuels**.[2] That includes heating our homes, schools, shops and other buildings. It includes burning fuel when we travel in cars, planes and other vehicles, unless they are fuelled by electricity produced sustainably. It also includes **fossil fuels** burnt to grow, transport, prepare and cook our food or make other things we use. The more of us there are, the more we use. How can we stop burning so much **fossil fuel**?

Individual changes

What lifestyle changes do you think individuals can make to have the biggest impact in reducing climate change?

Four major lifestyle changes would make a lot of difference: fly less, use cars less, eat a plant-based diet and have fewer children. The changes are not equal. For instance, having one less child would reduce **greenhouse gases** 75 times more than adopting a vegan diet (and 600 times more than upgrading household lightbulbs).[3] If you have taken some of the small steps in chapter 10, consider how much more impact you could have with lifestyle changes, such as taking one fewer flight a year, changing your car use, eating less meat or limiting how many children you have.

We discussed the first three changes in previous chapters. What about the fourth? How would you respond to someone who says, 'Children are a blessing from the Lord. God told Noah to fill the world with his offspring. So we should have lots of children'?

Our response:
Instructions to populate the earth had a different meaning when its human population was apparently eight (Genesis 9:7) than when it is approaching eight billion. A desire for a child may be biological and hard to resist, but raising a child uses a lot of resources and produces another person to consume

them. If you are thinking of having children in the future, could you plan to have one less child than you otherwise would have? Would you consider adopting a child, who already exists and needs love, instead of bringing another child into the world? Hundreds of people have joined a 'birth strike' – striking from having children for the sake of the planet![4]

It is natural to want to travel freely, eat anything we want and have as many children as we like. How much are these desires driven by greed? We covet what other people have. God told us not to. We do not have to give up everything, but let us reduce our **carbon footprints** as an act of love.

Beyond individual actions

Changing our lifestyle will help, but to reduce **greenhouse gases** more, we need to support interventions that can be carried out on a larger scale with many people. This brings us back to part of Figure 3, which we presented in chapter 2.

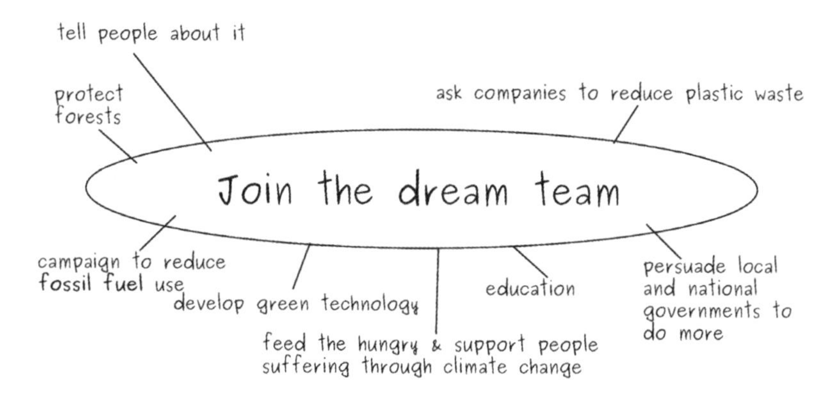

Figure 3 (detail)

Project Drawdown[5] measures the amount of carbon saved by different interventions. Many of these interventions require working together with others, and government policy changes, to have the most impact. The most effective ways of limiting temperature rise include onshore wind and solar farms, reducing food waste, eating plant-rich diets, education, family planning and reforesting tropical rainforests. Several of these are extremely cost-effective (having a large impact per amount spent).[6] We have discussed some of those interventions in other chapters. We will concentrate on a few others below.

Education

Increasing the number of years girls are in education reduces the risk of their having children at a very young age and marrying against their will. Education increases access to reproductive healthcare. Education empowers women economically, politically and socially, slowing population growth and reducing **carbon emissions**.[7]

Education also builds resilience and equips people to cope with the impacts of climate change. Education helps people to make choices to lower their **carbon footprint**.

Eleanor Hyde, BMS

A classroom in Guinea

Green energy and other green technology

With design improvements, energy can be produced more efficiently to provide heating, electricity and transport. Ideally, **fossil fuels** should now be left in the ground. To heat our buildings in a **sustainable** way, we will have to switch from **fossil fuels**, such as natural gas, to electricity generated from **sustainable** energy, like solar, wind and hydropower (being careful to avoid using them where they are not efficient).[8] That will not be enough, as there are always times when the sun doesn't shine, wind doesn't blow and water doesn't flow. Tidal power, heat pumps and geothermal power are more constant, but can only contribute a small proportion of our total energy. **Biofuels** (growing crops or trees to burn) may add to **carbon emissions**.[9]

For now, controversially, the best way of providing a constant carbon-free energy supply may be nuclear power.

Is nuclear power good or bad?

Not many environmentalists like nuclear power. Nuclear power has often been developed alongside nuclear weapons that can destroy civilisation in minutes. The consequences of a nuclear accident can be so devastating that they last for generations. The waste from nuclear power remains radioactive and a threat to all living things for thousands of years. We bury nuclear waste deep underground in concrete to keep it safe, but how do we know it won't be accidentally or deliberately released hundreds of years later?

We don't. But we should put these risks in context. If we add together all the risks of other sources of fuel, nuclear power turns out to be much safer. It makes no more sense to lump it together with nuclear weapons than it does to tie those in with X-rays and other types of nuclear medicine. Nuclear accidents are far rarer than accidents caused by other fuel sources, and on average less deadly. Producing oil kills about 20 times more people a year than nuclear power. Even wind power kills more people than nuclear power. Roads kill over 1,200 times more people than nuclear power.

What about waste? The non-nuclear industry leaves far more toxic waste (such as plastic, chemicals and even radioactive waste) strewn about the earth. One study suggested coal-fired power stations expose people to more radioactive waste than nuclear power stations do. Nuclear waste does not take up a lot of space in comparison. The toxic waste produced by Britain's nuclear power stations throughout their lifetime would take up one-tenth of a square kilometre, buried in a layer one metre deep.[10]

It may be possible to develop more constant carbon-free energy, from hydrogen, nuclear fusion or other new technologies.[11] We need more people to develop such technologies, including making progress with low-carbon innovation.[12]

If we take too long to reduce global **carbon emissions**, we will need to rely more on new technologies to reflect sunlight or *remove* **carbon dioxide**, not just to produce less. Proposed solutions include putting mirrors in space, injecting chemicals into the atmosphere, fertilising the ocean with iron filings, weathering rocks and burying charcoal underground. Technologies which do enough to solve the problem have not yet been developed and have unknown drawbacks.[13] We will need many new technologies to help us without creating further problems. For more detail about technological solutions, see **drawdown.org**, **es.catapult.org.uk/reports/innovating-to-net-zero** and **cat.org.uk**.

What young people are doing

Jo Rand

Martha Rand, 17, says, 'I have felt called to use my passion for science to aim for a career connected to climate change. I believe this is through engineering to find green solutions in areas such as energy.'

She explains, 'Throughout secondary school there were two things that stuck with me: wanting to change the world, and loving maths and science. For a while I didn't know how these two could link, but in year 10 I found my calling. We were learning about renewable energy in physics, and I just couldn't understand why all these innovative, environmentally friendly energy sources weren't being used more widely. For example, we have so much water in and around the UK, so why aren't we using it for tidal, wave and hydro energy? I decided that was what I wanted to do with my life. Over the past two years, this plan has evolved and developed, as I've learnt more about the crisis God's wonderful creation is facing, and how I might best use my talents to help it. I am planning to apply to Cambridge University, among other places, to study engineering. I love the idea of being an engineer, because it perfectly links the two things I've always known I wanted to do. To me, engineering is about applying maths, science and technology in order to improve the world around us.'

Faithfulness: working with others for large-scale change

In the New Testament, the words 'faithfulness', 'faith' and 'belief' come from the Greek *pistos*, which has the meaning of trust, solidarity and being on the same side. Being faithful to God means persistently obeying his commands and standing in solidarity with others who do.

How can we stand with other people to change energy use, develop green technology and improve education?

Jesus said, 'Blessed are the meek, for they will inherit the earth' (Matthew 5:5). He was alluding to Psalm 37:11, which lists 'the meek' among four groups of people who would inherit the earth – or, perhaps more accurately in Hebrew, occupy the land. The word for 'meek' in that psalm has the sense of being poor, perhaps because you've had your land and rights taken away from you. What about all the oppressed poor who have had their land stolen and exploited to extract minerals for rechargeable batteries and electronic components? Or whose land has been devastated by climate change? Is Jesus calling his church to bless the meek by speaking up for them and challenging the systems that keep them poor?

Persuade governments to act

We mentioned previously how your household can change its electricity supplier so that more of the profit goes towards **sustainable** power. But unless you generate your own electricity, you do not have much control over how it is generated. Electricity from a variety of sources goes into the national grid and is delivered to you. Governments have control over how their country's electricity is generated, by the way they manage the energy industry. In democracies, citizens influence the government's energy policy with their votes.

Energy use is just one of many aspects of climate change that governments have great influence over. Governments can also change how citizens and business are taxed and persuaded (by law or financial incentives) to make choices which benefit the **environment**. Governments can make it easier to use public transport, insulate our homes, farm our land, stop extracting **fossil fuels** and stop producing building materials like cement (which accounts for 8% of global **carbon emissions**).[14]

To slow down climate change and the poverty and oppression it causes, governments need to act together. What we do as individuals and families, or even communities and single countries, will help, but it won't be enough. Without large-scale international action, climate change will not slow down enough to avoid catastrophic temperature rises. Governments have been trying and failing to keep to targets (see Appendix A). We need to hold them to account.

What can we do to influence our governments? Sign petitions and write to your Member of Parliament or other leaders to put pressure on them to act. If you have a vote, or will one day, think carefully about how you will use it. Consider candidates' and parties' policies and track records.[15]

Influence other groups

National governments are not the only collectives that make a difference. So can multinational companies, businesses, local governments, churches, schools and clubs. Consider, for example, that 100 **fossil fuel** producers were responsible for 71% of **greenhouse gas emissions** between 1988 and 2015.[16] BP has set an ambition to become a **net zero** company by 2050 or sooner and to help the world achieve **net zero**.[17] How can we use our influence to create further change?

Should we campaign for banks to stop investing in fossil fuels?

If we want to keep our hands away from 'dirty money', we might choose not to use banks which invest in **fossil fuels**. If we want to go further, we might campaign for our church, denomination or bank to stop investing in **fossil fuels**.

That might help keep our choices ethical and raise awareness of the problem, but it doesn't necessarily stop **fossil fuel** funding. Investors who don't care about the climate impact can then buy up the investments at a lower price and spend their profits in harmful ways. If campaigns succeed in getting international oil companies (IOCs) to reduce their **fossil fuel** production, national oil companies (state-owned companies such as Saudi Aramco) will step in and increase their production. National oil companies are less transparent and less scrutinised than IOCs, and have a bigger average **carbon footprint** per unit of fuel produced, meaning that **carbon emissions** could rise.

All that might change once **fossil fuel** use plummets and becomes such a poor investment that banks and other institutions start withdrawing their investments on a large scale. Until then, more effective strategies might be to campaign for governments to tax **fossil fuel** use, eliminate **fossil fuel** subsidies, impose a tax on carbon, subsidise renewable energy sources and build a clean energy economy.[18]

What young people are doing

A group of young people in Christian Aid's Prophetic Activist Scheme launched a campaign in 2020 to ask the British government to stop funding **fossil fuel** projects overseas, and instead to fund projects that are in line with the Sustainable Development Goals.

Civil disobedience

What if you cannot persuade influential people to act, and your vote makes no difference?

After trying voting, petitions, marches, letters and many other strategies, with nothing changing fast enough and catastrophic climate change on the horizon, the founders of Extinction Rebellion (XR)[19] decided that more drastic non-violent direct action was needed to force governments to listen (see chapter 7).

XR uses disruptive protests to draw attention to the urgency of climate change and persuade governments to take effective action. Consider how quickly governments changed from seeing Covid-19 as a remote threat to telling the truth about it and taking action that no one could have imagined. If enough people realised how quickly we must act to prevent disastrous climate change, would governments change their policies?

Minorities of activists have achieved major changes through history, such as abolishing slavery; gaining votes for women; and ending British rule in India, racial segregation in the USA and apartheid in South Africa. Were such changes anti-democratic or ahead of popular opinion? When we look back on those changes, it is hard to imagine why people in power resisted them. But most did. Even when governments changed policy, many of their citizens disagreed. Change happened because enough people became convinced it was necessary. It often happened because enough of the population became so convinced that things needed to change that they were prepared to engage in non-violent civil disobedience.

Civil disobedience has achieved enormous change where marches, voting and violence did not. It means causing a nuisance without harming anyone. Christians faced lions rather than renounce their faith. Suffragettes chained themselves to railings. Gandhi refused to pay salt tax. Rosa Parks sat where she was not allowed to on a bus. Martin Luther King Jr organised sit-ins. Members of XR have upset people, but so did Gandhi, civil rights activists, suffragettes and Jesus. Civil disobedience is not meant to make everyone happy. It is meant to make people change their minds – what Jesus called 'repentance'.

Already by 2019, after a few months of civil disobedience, XR achieved changes. The UK Parliament declared a climate emergency, set a **net zero carbon emissions** target for 2050 and set up a citizens' assembly with a limited remit. Made up of a representative group of ordinary citizens, the assembly listened to evidence about climate change and published recommendations for the government in September 2020.[20] All of these were steps forward, but still many steps away from the urgent action that both XR and the **IPCC** were calling for. We need government policies on an international scale around transport, how we get our energy, how we build and how we protect forests and wildlife.

There are many other ways you can persuade people to rethink. Instead of getting involved with XR this year, we chose to put time into writing this book. Sarah Corbett promotes the art of gentle protest. She speaks about how anyone can make hand-crafted gifts for influential people, which can be surprisingly good at persuading them to make fair decisions that improve many lives.[21]

Matthew 5 and non-violent civil disobedience

Jesus' instruction 'If anyone slaps you on the right cheek, turn to them the other cheek also' (Matthew 5:39) suggests a kind of civil disobedience. The person who offers the other cheek regains some power in the situation. It is not the same as just letting someone hit you again and again. It forces the other person to think about the bad things they are doing, and maybe decide not to carry on with them.

It is similar for Jesus' other examples. Jesus said if someone wants to sue you for your inner garment, give them your outer garment as well (v. 40). At the time, most people only had these two garments. Doing this would leave them naked and demonstrate the shamefulness of such an unjust system. Those seeing the naked person would be the ones who were shamed. Jesus also said, 'If anyone forces you to go one mile, go with them two miles' (v. 41). Soldiers were allowed to force someone to carry their belongings for a mile, but not for longer. Insisting on going a second mile would make the soldier uncomfortable as they would be breaking military code. The powerless person would become powerful.[22]

God of power and might,
we believe that nothing is impossible for you. Please guide us and give us wisdom so that we might be effective in our actions. Help us not to limit what can be done by thinking about how small we are. Help us to unite and work with others to make this world a better place. Amen

Loving our neighbours worldwide: Coalition for Rainforest Nations

rainforestcoalition.org

One of the most cost-effective charities dealing with climate change is the Coalition for Rainforest Nations (CfRN).[23] It works by persuading countries to change their national policy and keep the forests they have.

The CfRN began because of the international failure to deal with climate change.[24] Many countries with tropical rainforests agreed that it would be better not to cut them down, but did not want to handicap their development. CfRN persuaded richer countries to pay poorer countries to keep their rainforests. That way the world would benefit from not losing rainforests, without slowing development. Without CfRN, these agreements would probably have taken much longer or not happened at all.

There is much work still for CfRN to do. For instance, two major rainforest nations are not members – Malaysia has never joined and Brazil recently left and has resumed destruction of the Amazon rainforest. Many nations still need to produce a national plan for protecting their rainforest. Nevertheless, CfRN is thought to have been extraordinarily effective, saving a tonne of **carbon emissions** for every few cents spent, far more than most **carbon offsetting** schemes.

Young people around the world: Melting ice and coastal erosion in Alaska

Robert Knoth/Greenpeace

Gabriel Stenek, 17, is from Shishmaref, on Sarichef Island in Alaska. Gabriel is faced daily with the effects of climate change. Storm surges have become more frequent and dangerous, and cause coastal erosion and the loss of land. Rising temperatures have resulted in a reduction in the ice that buffers Shishmaref from storm surges. The **permafrost** that the village is built on has also begun to melt. At least 18 houses have fallen into the sea; the photograph above shows one of them. The 600 or so people of Shishmaref have voted to move to the mainland. Leaving the place where they have always lived is painful, but they were losing so much land that they could no longer sustain themselves there. This has given Gabriel a passion to warn others about climate change, and so he is an Arctic Youth Ambassador.[25]

Jamie's tips

1. Pray for technological solutions.

2. Look at Shishmaref on Google Maps and see the photos which have no snow, despite how far north it is. Think about how Gabriel and his family feel, knowing that they have had to leave their island because of climate change. Look at the photo of the house falling into the sea. Think about how you would feel if that was your house.

3 Read more Bible verses on justice. Start with Deuteronomy 32:4; Psalm 82:3; 89:14; Isaiah 58:6–7; Amos 5:24; Micah 6:8. Add any others you think of.

4 Read more stories from young people around the world whose lives have been affected by climate change, at **ourclimatevoices.org** or **srhr-ask-us.org/publication/we-stand-as-one-children-young-people-and-climate-change**. Or search for 'climate impacts' at **climatevisuals.org**.

5 Find out more about the Coalition for Rainforest Nations. See **rainforestcoalition.org**.

6 Look at the climate policies of different political parties, so that you can take these into consideration when you have a chance to vote.

7 Eat less red meat, use cars less and fly less. I have said this before, but these are individual changes that make a big impact, so they are worth repeating!

8 Whatever your age, find someone to mentor. Help them study the Bible and make lifestyle changes to care for creation. Maybe study this book with them.

9 Spread the news about climate change.

10 Read about what other countries have done to become more **sustainable**.[26] Then contact your MP or the leader of a political party and ask them to take action for the climate (see **hftf.org.uk**).

11 Attend legal, non-violent **climate justice** protests or strikes. Find out more about XR or their youth wing, XR Youth. Or find out about gentle protest at **craftivist-collective.com**.

12 Martha has used her passion for the **environment** to guide her career choice. Consider whether your concern about climate change could influence your work or your volunteering. Volunteer with one of the charities in this book,

fundraise for them or do work experience or a gap year with one of them. Or bring your care for the **environment** into your work or studies.

Which of these do you already do?

____

____

____

Which will you attempt to do this month?

____

____

____

12 Join the dream team

Bible passages

1 CORINTHIANS 12:4–6, 12–28

There are different kinds of gifts, but the same Spirit distributes them. There are different kinds of service, but the same Lord. There are different kinds of working, but in all of them and in everyone it is the same God at work…

Just as a body, though one, has many parts, but all its many parts form one body, so it is with Christ. For we were all baptised by one Spirit so as to form one body – whether Jews or Gentiles, slave or free – and we were all given the one Spirit to drink. And so the body is not made up of one part but of many.

Aaron Lewin/Wycliffe Bible Translators

Now if the foot should say, 'Because I am not a hand, I do not belong to the body,' it would not for that reason stop being part of the body. And if the ear should say, 'Because I am not an eye, I do not belong to the body,' it would not for that reason stop being part of the body. If the whole body were an eye, where would the sense of hearing be? If the whole body were an ear, where would the sense of smell be? But in fact God has placed the parts in the body, every one of them, just as he wanted them to be. If they were all one part, where would the body be? As it is, there are many parts, but one body.

The eye cannot say to the hand, 'I don't need you!' And the head cannot say to the feet, 'I don't need you!' On the contrary, those parts of the body that seem to be weaker are indispensable, and the parts that we think are less honourable we treat with special honour. And the parts that are unpresentable are treated with special modesty, while our presentable parts need no special treatment. But God has put the body together, giving greater honour to the parts that lacked it, so that there should be no division in the body, but that its parts should have equal concern for each other. If one part suffers, every part suffers with it; if one part is honoured, every part rejoices with it.

Now you are the body of Christ, and each one of you is a part of it. And God has placed in the church first of all apostles, second prophets, third teachers, then miracles, then gifts of healing, of helping, of guidance, and of different kinds of tongues.

MATTHEW 5:9

'Blessed are the peacemakers, for they will be called children of God.'

What does the passage from 1 Corinthians say to you? Which words or verses stand out to you?

____

____

____

____

____

One body

It is wonderful that we can all join together as members of one body, all across the world, and be part of a dream team trying to obey God in all aspects of life – including caring for the world he created and showing love to others by being considerate about how we tread on the earth.

We are thankful for all the members of the dream team we have met in this book. They bring many different gifts. There are teachers and those with gifts of healing, helping or guidance (v. 28). There are humanitarian workers, health workers, engineers, photographers, translators, cooks who feed the hungry, administrators, scientists and people who care and pray. Young people strike, write letters, plant, give talks and pick up litter.

We could say:

> Now if the photographer should say, 'Because I am not a teacher, I do not belong to the team,' he would not for that reason stop being part of the team. And if the engineer should say, 'Because I am not a nurse, I do not belong to the team,' she would not for that reason stop being part of the team. If the whole team were nurses, who would construct the water tanks? If the whole team were teachers, who would take the photographs to raise awareness? But in fact, God has placed us, every one of us, just as he wants us to be. If we were all the same, how would we be able to achieve the many different tasks which need to be done? As it is, we are all different, but we are all one team.

As well as helping in the response to climate change and disasters, the mission organisations featured in this book support workers who teach the big story of the Bible. There are people who plant churches as well as those who plant trees. They are all part of the team.

Do you want to be part of the team? Which part will you play? Reducing your carbon footprint; praying; protesting; telling others; fundraising; encouraging those who are making a difference; helping develop green technologies – these are just some of the ways to be involved in the team.

____

____

____

____

____

Mentoring others

Nuevas Esperanzas

Teamwork in Nicaragua

Parents, teachers, youth workers and those who work with children in churches have an important role in the team. They can encourage and enable young people to act for **climate justice**. We can all offer to mentor others.

What young people are doing (with support from teachers)

Ruth Dockree, 10, says:

Sarah Dockree

Ruth and her brother Samuel at a protest

I wanted to make a difference, so I went up to my school's deputy head teacher and expressed my concern about the climate crisis. This led to us having two weeks of spreading environmental awareness at school. I spoke in assembly and we held a competition to design the best realistic machine that could tackle climate change. We also had a 'no single-use plastic day', when we encouraged all the children and staff to go through the day without using single-use plastics.

> We even spoke to a local councillor about what we were doing when they came into school. During our school's Culture Week, when we celebrate different countries and cultures, I had the opportunity to speak in assembly about the impact of climate change around the world. All of this happened because I shared my concerns with my teacher, who was really excited about working together to make things better.

If one part suffers, every part suffers

The apostle Paul wrote that 'there should be no division in the body', but 'its parts should have equal concern for each other. If one part suffers, every part suffers with it; if one part is honoured, every part rejoices with it' (1 Corinthians 12:25–26).

The body is suffering from the impact of climate change. The lungs are infected with pollution. The kidneys are producing kidney stones through lack of water. The stomach is empty. The feet have been burned, running through wildfire. An arm has been broken by a falling tree. We are one body, and the body is in pain. Can you feel the suffering of the parts of the body?

Many people around the world are suffering the impact of the changing climate, which has left them hungry, thirsty, sick and homeless. Many more will suffer in the future. We are called to suffer with them (v. 26). Our suffering may look different. We may suffer by choosing to deny ourselves some of the things we would like (e.g. flying, gas-guzzling cars, meat or having children) to support our brothers and sisters. We can choose to donate money to help others, instead of buying more stuff for ourselves.

1 Corinthians 12 teaches us that we are all the same, regardless of our race, 'whether Jews or Gentiles, slave or free' (v. 13). Paul expands this idea in Galatians 3:28: 'There is neither Jew nor Gentile, neither slave nor free, nor is there male and female, for you are all one in Christ Jesus.' In this book we have met people from every continent that is home to humans. We have met people with and without disabilities, from different races, ages and stages of life. There is no 'them and us'. We are not divided into 'victims' and 'helpers'. We are all affected by climate change, and we all suffer because we care about those who are suffering. And we can all play a part.

What young people are doing

A reminder of some of the global team we have met:

Think of Janet from Kenya, whose family suffered health problems due to **fossil fuels** and who is now a young activist working for climate justice. Or Oliver in Peru, who suffers from the weather and pollution and takes action by planting trees and clearing litter. Or Grace in Burundi, who was abandoned at birth but now helps others out of poverty. Or Bhim Maya in Nepal, who was in an earthquake and now wants to provide education. Or Surbana in Bangladesh, whose school was flooded and who wants to help with legal issues. Or Gabriel in Alaska, who is an Arctic Youth Ambassador, spreading awareness. We are all in the same team, empowered to do what we can as we work together.

Blessed are the peacemakers

Is this dream team a peacemaking team? What has peacemaking got to do with dealing with climate change?

In Hebrew, peace is *shalom*. It means a lot more than just putting an end to war. *Shalom* is about wholeness, well-being and contentment for everyone. It means that everything wrong is put right. Think back to Figure 3 in chapter 2 (p. 37). If God can use us to stop the decay in creation, he will make us peacemakers. Jesus said we would be called children of God (Matthew 5:9).

There must be more than this

Perhaps you are thinking, 'There are other important things for Christians to do.' You would be correct. We are not trying to cover everything here. Other 'parts of the body' have written excellent books on evangelism, discipleship, prayer, worship, other aspects of the **ecological** crisis and all manner of other topics. We are glad they have. There is much more to the Christian life than what is discussed in these pages, and we pray that you will explore the whole of the Bible. But that does not let us off the hook of obeying God in the ways covered in this book.

We are pleased to have this opportunity to share about a topic that is not often spoken about in churches or discussed in Bible studies. If this was taught more, we would not need to write about it. While we were writing this book, a Christian friend said, 'I do not share the same level of concern or feel that climate change is a reason for [hundreds of Christians] not to fly across the world for a short conference.'

Until a few years ago, we also regularly flew to mission conferences without thinking about the environmental impact. Then Jamie started to speak about climate change. We looked at the science and the Bible, and our views and behaviour changed. This is why we now feel compelled to put our gifts to use by writing, passing on what we have learned.

There is much more to climate change than we have been able to cover here. For further information, please see the resources in the Appendices.[1]

Don't be anxious

How many Bible verses can you think of that include phrases like, 'Do not worry', 'Do not be anxious', 'Fear not' or 'Be at peace'?

____

____

____

____

These phrases appear well over 100 times in the Bible. Rather than telling us off for worrying, they are generally words of reassurance, reminding us that we do not need to worry. God has got this.

Thinking about climate change can leave us feeling anxious. But Jesus said, 'Do not worry, saying, "What shall we eat?" or "What shall we drink?" or "What shall we wear?"... But seek first his kingdom and his righteousness, and all these things will be given to you as well' (Matthew 6:31, 33).

What do you think it means to 'seek first his kingdom and his righteousness'?

If we keep our focus on God, we do not need to worry. If we all play our part in the dream team, we will work together to help provide for those in need.

Paul wrote:

> At the present time your plenty will supply what they need, so that in turn their plenty will supply what you need. The goal is equality, as it is written: 'The one who gathered much did not have too much, and the one who gathered little did not have too little.'
> 2 CORINTHIANS 8:14–15

If we live in this way, we do not need to worry, as we will help each other. When we focus on helping others, we are less likely to worry about ourselves. If we all play our part in tackling climate change, together we can make a real difference.

John Wesley was credited with saying:

> *Do all the good you can,*
> *By all the means you can,*
> *In all the ways you can,*
> *In all the places you can,*
> *At all the times you can,*
> *To all the people you can,*
> *As long as ever you can.*[1]

If we all do this, we will live well, and we might save the planet at the same time. Let's dream together, and be a team together to make peace.

Now may the Lord of peace himself give you peace at all times and in every way. The Lord be with all of you.
2 THESSALONIANS 3:16

Loving our neighbours worldwide

We have listed some great charities in this book. There are many more we would have liked to mention. We invite you to support a charity of your choice that works to tackle the impact of climate change.

Jamie's tips

This time I simply challenge you to act on what you have read in these pages. *You* can be the final case study in this book.

What I am doing to help change the climate:

___

___

___

___

___

___

___

___

___

___

Contacting us

If you like, tell us what action you have taken because of this book (and perhaps you will feature in the next edition!). Use our contact form at **resilientexpat.co.uk/misc.php**. Feel free to also contact us if you think we have missed something or got things wrong.

You can let us know if you spotted every book of the Bible in this book. Please keep reading the Bible and finding out more. If you don't have a Bible of your own, we recommend getting a version which you find easy to understand.

To close, we have adapted Rudyard Kipling's poem 'If' for the current day.

If, in a climate crisis

If you can ride your bike when all about you
Are driving cars and showing them to you,
If you can trust yourself to reuse nappies
But not judge those who do not manage to;
If you can walk and not be tired by walking,
And for your holiday choose not to fly,
And practise love, and not give way to hating,
And never boast of what you do not buy:

If you can dream – but never stop at dreaming;
If you can think – and act upon your plan;
If you can do what must be done to save lives
And worry not what's said on Instagram;
If you can bear to hear the truths you've spoken
Turned to fake news by those who jeer at you
Or watch the things you gave your life to, broken,
And pick them up and start again anew:

If you can march with crowds and stay non-violent,
Or sit alone outside your parliament,
If you rejoice when carbon footprints reduce
But greet increases with sincere lament;
If you plant trees, and hope they grow to bear fruit,
And don't forget the things which matter most:
God's is the earth and everything that's in it,
Let's care for it, or we will all be toast!

Appendix A

Understanding climate change

> **There is no faithfulness, no love,**
> **no acknowledgement of God in the land...**
> **Because of this the land dries up,**
> **and all who live in it waste away;**
> **the beasts of the field, the birds in the sky**
> **and the fish in the sea are swept away.**
> HOSEA 4:1, 3

In this Appendix we summarise how the climate is changing and why; its current and future consequences; and other ways humans are harming the environment.

Weather isn't climate

First of all, weather is not the same as climate. Weather changes day to day and year to year. Climate is what the weather is generally like when it is averaged over years. We can't observe climate change in any one year. In Europe and North America there can still be cold winters with lots of snow. You might be surprised to know that there is more snow in winters just below freezing than when winter is much colder. Some decades are hotter and some colder. Those changes have happened all through the earth's history.

The world's average temperature has risen

But when we measure the changes over decades, we can see that, overall, the temperature is going up and up. Figure 4 shows how the world's average temperature has increased since 1900, and steeply since the 1970s.

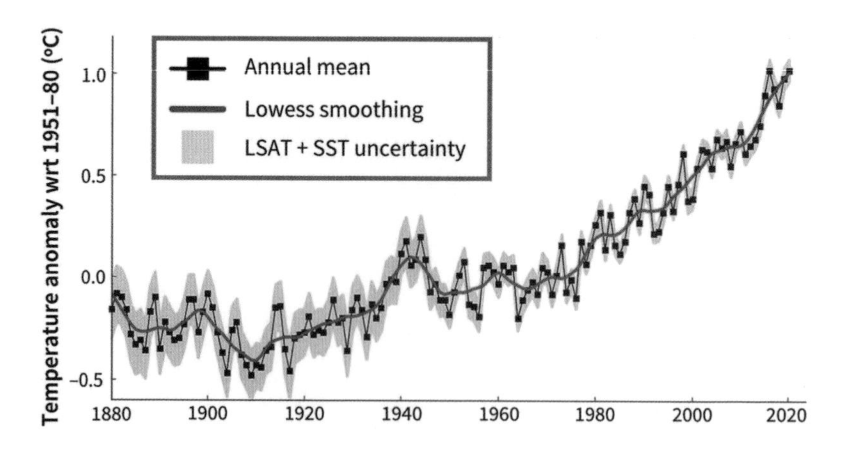

Figure 4 Global land-ocean temperature index[1]

The earth is about 1°C hotter than it was 150 years ago, when the Industrial Revolution really took off.

Why 1°C matters

1°C might not sound like much, but it is an average – a way of summarising a lot of bigger changes, such as:

- much greater temperature rises near the poles (average 3°C or more);
- ice-free Arctic summers;
- longer **droughts** in Africa;
- more extreme temperatures;
- greater warming on land than at sea;
- long, hot summers in Europe, sometimes hot enough to kill thousands;
- dangerous heatwaves in cities.

In the 17th century, a 1–2°C drop in world temperature was enough to cause the 'Little Ice Age'.

Extreme weather events

We are having more extreme weather events. Storms, hurricanes, cyclones, tornadoes and snowstorms are getting more intense and dangerous. Dry places are getting drier and wet places wetter. Parts of the world that rely on rain for crops to grow are getting shorter and heavier rainfall, with longer dry periods in between. This increases the risk of sudden heavy (flash) floods.

Milind Ruparel on Unsplash

That kind of pattern is no good for growing food crops. Flash floods drown animals and people, destroy homes and plants and bring deadly diseases. If the land then dries out for a long time, any remaining plants will die. Forests and fields turn into deserts.

Melting ice

Another major consequence of rising temperatures, and other climate-related changes, is melting snow and ice – especially as temperatures are rising more in the polar regions and snowy mountains than elsewhere. About a third of the world's population live in China and the Indian subcontinent, which get most of their water from the glaciers in the Himalayas. It is okay when ice melts slowly and feeds rivers, but ice melting quickly eventually leaves rivers dry and people thirsty. Ice is melting so fast that sea levels are rising and won't stop rising for centuries, even if the human race died out now.

Sea level rises

Between 1993 and 2016, Greenland lost an average of 286 billion tons of ice per year. The rate of ice loss in Antarctica has tripled in the last decade,[2] and is accelerating every year.[3]

Danting Zhu on Unsplash

The sea level has risen at least 16 cm since 1902 and is now accelerating at over 3 cm per decade. This might not sound like much, but the sea level rises and falls all the time with the tide. An overall sea level rise causes damage through higher tides, more floods, sea water destroying crops, animals losing **habitat** and coasts eroding. Millions of people live close to the sea, in large cities, tropical islands and river estuaries. The land they live on will disappear under the sea. Destruction will come when there is a storm surge or flood.

How people are affected

Sometimes it is possible to buy your way out of problems. Rich people, including most people in Europe and North America, can move somewhere else, protect their homes, buy enough food and water or rely on their governments to keep them healthy and fed. For example, the Thames Flood Barrier is expected to protect London from flooding until 2070.[4] Poorer people don't have those choices. The changes to climate and land use make it harder for people to get food from the land (by farming, foraging or hunting) or to earn enough money to live on. Particular groups of people, and parts of the world, are in more danger and get poorer as climate change increases.

What happens when people are forced out of their homes because their land disappears underwater or their homes keep being flooded? What about when people have to move because they don't have enough water, the land isn't giving them crops or feeding their animals or it is too hot to stay? What about when storms threaten their homes and diseases threaten their health? Or when for years they cannot find work or make enough money to eat and feed their families?

Many are dying already. People die because they do not have enough to drink or eat. People die in floods and storms, in heatwaves and pandemics, and because they are poor. People always have. But as the planetary crisis gets worse, more people die earlier than they would have.

People try their best not to die. We have two natural responses to danger: fight or flight. Often people end up fighting each other. Brutal and long civil wars are dressed up in politics, but they often have a lot to do with fighting over limited resources like water. The devastating Syrian civil war came about partly because of **droughts** and crop failures.[5]

Or people flee. It takes a lot to leave your home and everything you know, especially if you have to travel far away and have no idea how you'll manage it. But if you are going to die otherwise, it's a pretty good option. Most move within their country or go to a nearby country, but some spend years trying to get to somewhere far less affected by climate change. People often call them economic migrants and say they should go back home, as if climate change were not putting their lives in danger. As countries near and far protect their borders and turn migrants away, migrants continue to die in the attempt to break through, drowning in the sea or suffocating in refrigerated trucks. It is estimated that 1.2 billion people are at risk of becoming **climate refugees** by 2050.[6] As living conditions worsen in Africa and elsewhere, it is not hard to see how the pressure of migrants on the move could lead to more violence and war.

Let us now think more about why the earth is warming.

The greenhouse effect

It starts with **greenhouse gases** – gases like **carbon dioxide**, sulphur dioxide, nitrous oxide, water vapour and **methane**. **Greenhouse gases** are not all bad. Natural processes produce and absorb them. Without them there would be no life on earth. Plants use **carbon dioxide** and sunlight to make energy so they can grow and produce oxygen that animals need to breathe. Breathing changes oxygen into energy and **carbon dioxide**. **Carbon dioxide** and other

greenhouse gases let sunlight through to the earth but do not let the infrared radiation from the earth back out, instead absorbing it. It's a bit like glass letting the sun into a greenhouse and then trapping heat inside, making the greenhouse warmer. Because of this greenhouse effect, the earth stays warm enough for plants and animals to live and grow. If plants absorb all the **carbon dioxide** produced on earth, **carbon dioxide** levels don't change and all is fine.[7]

Carbon emissions (**carbon dioxide** entering the air) become a problem when there are too many of them, trapping more heat in the earth than it needs. For hundreds of thousands of years the amount of **carbon dioxide** in the air has varied from 170 to 300 parts per million (ppm), before shooting up rapidly during the Industrial Revolution, to now over 400 ppm (see Figure 5).

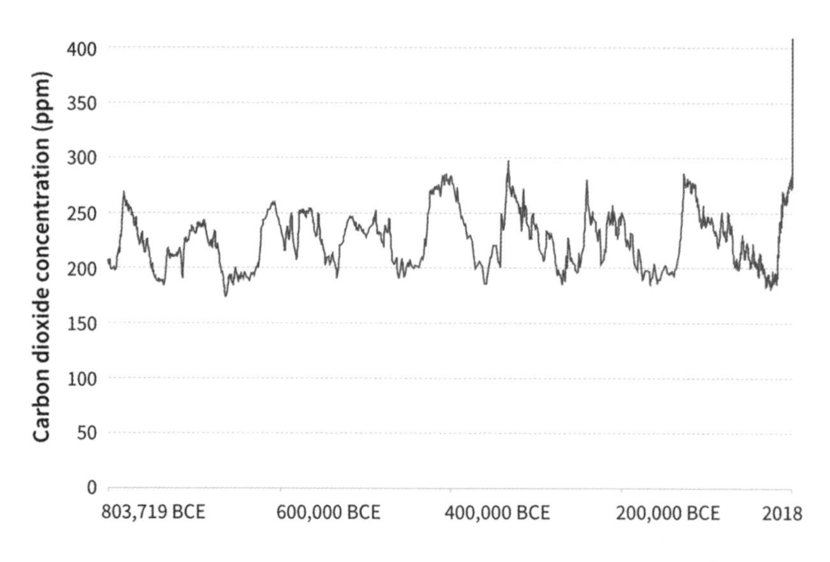

Figure 5 Global average long-term atmospheric concentration of CO_2[8]

Greenhouse gas emissions caused by humans

Carbon emissions are not caused just by animals breathing and other natural processes, but are also made by humans. **Carbon dioxide** enters the air whenever we burn **fossil fuels** (coal, wood, natural gas, petrol and other oil products) or paper, trees, **biofuels** or plastics – in fact anything **organic**. Most of the energy we use for heating, electricity, transport and making things comes from burning **fossil fuels** and releases **carbon dioxide**. The Industrial

Revolution, population growth and consumer demand has caused a very rapid increase in how much we are burning these.

Burning some **fossil fuels** (like coal, wood and diesel fuel) also releases sulphur dioxide, nitrogen oxides and carbon (soot). These make up less of the atmosphere and don't stay as long in it as **carbon dioxide**, but they trap much more heat, so the temperature keeps rising.

Industrial processes produce nitrous oxide and sulphur dioxide, and farming (especially when it is intensive or uses nitrate fertilisers) produces a lot of nitrous oxide. Grazing animals like cows produce **methane**, as does rice in paddy fields. Sulphur dioxide, nitrous oxide and carbon make it harder to breathe when they reach large concentrations, such as in city air, and can kill people and animals.

It is not just in the air. Gases like sulphur dioxide react with rainwater and make it acidic, so that it kills plants and forests. Soot settling on snow and ice darkens the earth, so that it absorbs more sunlight and warms further. The oceans absorb a lot of **carbon dioxide** and become more acidic too with less oxygen, making it harder for fish and sea animals to live there. Coral reefs like the Great Barrier Reef off Australia – once a wonderful place to see beautiful sea creatures – have almost completely died.

Did you know?

Scientists are predicting that coral reefs will disappear this century due to climate change if we do not act. This is important because coral reefs have a high biodiversity of creatures in them, and directly support 500 million people. Also, they protect land from storms and tidal waves.[9]

How else are humans harming the environment?

Climate change is one of several features of what geologists call the Anthropocene – the era in which humans have had a significant impact on the **environment**. Both climate change and other human influences are deadly for wildlife. Animals and plants die because of changes in the seasons – summers that are too hot, winters not cold enough – and the damage brought about by extreme weather. In 2010, the UN set 20 targets for reducing **biodiversity** loss by 2020. Not one of those targets was achieved.[10] Since 1970 the world has lost more than two-thirds of its population of mammals, birds, reptiles, amphibians and fish.[11] Insects are disappearing even faster.[12] About eight species of plant have disappeared every three years since 1900 – up to 500 times faster than plants normally go extinct. This matters even if you don't care much about plants or animals, because they are part of the **biodiversity** of **ecosystems. Ecosystems** are how plants and animals live together in a particular **habitat** (like a meadow, a tropical rainforest or a rocky seashore) and keep each other going. One species can be vital to the way the others live. Losing one species can destroy a whole **ecosystem**, including any food or other materials that humans or other living things need to stay alive.

Did you know?

Maybe you don't like bees. They sting. But as well as making honey, bees spread pollen between flowers. Without bees (and other insects that pollinate flowers, including wasps and mosquitoes), many plants could not reproduce. Many ecosystems would be damaged and other plants and animals would be lost. It would become much harder to grow fruit and vegetables. They would be far less available, less varied and more expensive. Human nutrition would suffer. We still have bees, but their populations are in decline all around the world, mainly for reasons related to climate change and human effects on the environment. If the bees die out, we are in deep trouble.[13]

Verna Langrell

Loss of wildlife is caused not only by climate change but also by overfishing, overgrazing and **industrial farming.** To feed humanity at as low a price as possible, we trawl huge fishing nets to catch more fish than we need; we destroy **ecosystems** to graze animals for meat or we raise animals in conditions that spread disease; and we grow crops on an industrial scale that pollutes and wears out the land, sometimes until it turns into desert.

We cut down forests not only for farming, but also to build cities, factories and mines. After forests are cut down, many of these activities produce **greenhouse gases**, with fewer trees around to absorb them. It gets worse when we deliberately burn forests to clear land, or the land becomes so dry that wildfires start quickly and are put out slowly. Or when we lay paving and remove hedges and terraces from farms, which makes it harder for the land to drain and easier for it to flood.

Animals and plants we have transferred across continents have spread rapidly and destroyed **ecosystems** and **biodiversity**. Mining has destroyed forests and other **ecosystems**, polluted rivers and displaced or enslaved local people. Large hydroelectric dams are depriving river deltas of sediment, so they cannot keep up with rising sea levels and they dry up.

Deadly diseases now spread more easily from wild animals to humans. Some (like malaria) spread more as the weather gets hotter. Others (like cholera) spread in heavy floods. Epidemics like flu, plague, Zika, Ebola, HIV and Covid-19 have spread because humans came into closer contact with wildlife, for instance through road building, **industrial farming** or craving wild meat.

What is likely to happen next?

All this damage is already happening. But what are we heading for? No one can know the future, but looking at the way nature and the climate have been going allows us to make reasonable predictions. Scientists around the world have joined to form the **IPCC**. In 2018 the **IPCC** produced a detailed report looking at what is likely to happen up to the year 2100 and beyond, depending on what we do about it.

The **IPCC** showed how much the climate has been changing, with the world's **carbon emissions** tending to increase every year. They said that if we could start reducing **carbon emissions** by 2020, we would stand a chance of limiting the average temperature rise to 1.5°C (0.5°C higher than it has already risen) by 2100. We could do that by achieving **net zero carbon emissions** by 2050 – where any **carbon emissions** would be balanced by methods of absorbing carbon back into the earth (like having enough trees planted to absorb them). If we achieve that, the average temperature would probably rise higher than 1.5°C for a few decades, before settling back at 1.5°C by the end of the century.

The changes we have already described would keep on happening and worsen if we manage to limit the world to a 1.5°C rise. But they would not be anything like as bad as a 2°C rise. At 1.5°C we could avoid some of the worst consequences of rising sea levels, uninhabitable land and mass migration. There is a one in five chance that we will at least temporarily reach 1.5°C by 2024.[14] To stabilise at 1.5°C, every year beyond 2020 that we fail to reduce **carbon emissions** will give us two fewer years before we have to reach **net zero**. For instance, if we don't start reducing **carbon emissions** till 2025, we would have until 2040 to reach **net zero**. If we don't start until 2030, it will be too late.

Are we making progress towards the changes the **IPCC** recommended? Hardly. Even in 2020, with a temporary reduction in **carbon emissions**, the **carbon dioxide** in the atmosphere still rose. It's the total amount that counts, and that will keep rising until we reach **net zero**.

Some governments are making more changes than others. But governments have to agree together to take action, and stick to their agreement, for us to get to **net zero**. Governments have been talking for decades and have still failed to reach an agreement after the **IPCC** report in 2018. The USA decided to leave the Paris Agreement, a 2015 international agreement that aimed for a 2°C rise by 2050. China, which has the highest **carbon emissions** of any country, plans to start reducing them in 2030 and to reach **net zero** by 2060. The UK government continued to praise itself for reducing **carbon emissions** and called for China to reduce its own, without counting how many of China's emissions are produced because British people are buying goods from China. Governments disagreeing will not get us to **net zero** when worldwide **carbon emissions** continue to increase.

If the world leaves serious action on climate change longer, we will have to take even more drastic steps to reduce **carbon emissions**, or rely on technology that hasn't been developed yet and has its own risks. All the problems of climate change we have described will get much worse.

What are feedback loops?

Feedback loops occur where one problem causes a second problem that makes the first problem worse and makes the planet warm faster. Here are three examples.

1. Wildfires: a warmer climate dries out vegetation and makes it harder to stop wildfires; these fires then add more **greenhouse gases** to the atmosphere.
2. Melting ice: ice reflects 90% of sunlight back out into space, keeping the earth cooler. But as ice melts, white snow turns to clear water, which absorbs 94% of the heat from the sun (that is why it looks blue) instead of reflecting it away. So the sea warms, and so more water evaporates into the air as water vapour. Water vapour is naturally in the air but also acts as a **greenhouse gas**, so more water vapour keeps more heat in.
3. **Permafrost**: this is the name for land near the ice caps which has been permanently frozen. As **permafrost** begins to melt, **methane** underneath it will be released into the air and speed up global warming, which in turn will melt more ice and raise sea levels by metres, not centimetres. There are signs that this has already begun.[15]

If we carry on as we did up to the end of 2019, we are on course for a 3–4°C rise by 2100. If we follow the path of highest emissions, we could reach 4°C by 2070 and 7°C by 2100.[16] These are average temperatures, which summarise much greater changes. The temperature rise will be even greater if feedback loops operate, like the release of **methane** from **permafrost** (see above). A rise of 3°C or above is so catastrophic that the **IPCC** barely considered it in 2018. It could mean it would be impossible to live in large parts of the world, and many millions of people would die. Those who survive could be fighting over land and food in the northern parts of Russia, Canada and Scandinavia.

Our planet has survived great crises before. The Black Death possibly killed more than half of Europe. Genesis 7 tells of a flood which wiped out most of life on earth. Life on earth survived. Life as we know it may not. Carrying on as we are would be like sentencing millions of people to premature death.

We can't stop all changes to the climate and the **environment**. But we can slow them down. The British government's former chief scientific adviser Lord May suggested that faith communities could do what scientists and politicians could not, and motivate people to change their behaviour. Have we succeeded?

In 2020, the world changed many things quickly to protect people from Covid-19. What if we treated that as a warning of worse to come or a practice run for dealing with it?

The **IPCC** said to achieve **net zero** we need to make drastic changes to the way we live. What are *you* going to do to change the climate?

Appendix B

Carbon emissions and climate/food vulnerability for different countries

In 2018 the European Commission produced a report entitled *Fossil CO_2 Emissions of All World Countries*.[1] We recommend this if you wish to study **carbon dioxide** emissions of all 209 countries listed (although we disagree with their definition of some countries).

In 2019, Christian Aid used this information, combined with data on vulnerability to climate and food insecurity, to produce the report *Hunger Strike: The climate and food vulnerability index*.[2] This lists 113 countries for which full data were available. The data in tables B1 and B2 are taken from that report and used with permission from Christian Aid.

Table B1 lists **carbon emissions** but not how much carbon is reabsorbed. Bhutan is the first **carbon negative** country in the world. It emits about 1.8 tonnes of **carbon dioxide** per person, but because over 70% of the land is covered in trees, the forest absorbs about three times more than the amount of carbon emitted.[3]

Table B1 Countries most vulnerable to climate change and food insecurity (out of 113 studied)

Climate and food vulnerability index rank (1 = most vulnerable)	Country	Food insecurity score (a lower score means greater insecurity of food)	Carbon dioxide emissions per person (tonnes)	Carbon dioxide emissions per person rank (1 emits the most, 113 the least)
1	Burundi	23.9	0.027	113
2	Democratic Republic of Congo	26.1	0.043	112
3	Madagascar	27	0.163	104
4	Yemen	28.5	0.443	90
5	Sierra Leonne	29.2	0.173	103
6	Chad	31.5	0.062	110
7	Malawi	32.4	0.084	109
8	Haiti	33	0.321	96
9 (equal)	Niger	33.7	0.116	107
9 (equal)	Zambia	33.7	0.29	97

Table B2 Countries that contribute the most carbon emissions per person (out of 113 studied)

Carbon dioxide emissions per person rank (1 emits the most, 113 the least)	Country	Food insecurity score (a lower score means greater insecurity of food)	Carbon dioxide emissions per person (tonnes)	Climate and food vulnerability index rank (1 = most vulnerable)
1	Qatar	76.5	37.052	92
2	Bahrain	67.8	23.968	73
3	Kuwait	74.8	23.486	86
4	United Arab Emirates	72.5	21.574	83
5	Saudi Arabia	72.4	19.393	82
6	Oman	74.4	16.915	85
7	Canada	83.2	16.855	105
8	Australia	83.7	16.452	108
9	USA	85	15.741	110
10	Kazakhstan	57.7	14.623	57
…				
37	UK	85	5.729	110

There are also other ways of assessing which countries are most vulnerable to climate change. The Global Climate Risk Index[4] is a separate measure that analyses fatalities as well as economic losses that occur due to extreme weather events. The results differ from year to year, but like tables B1 and B2, they indicate that less developed countries are generally more affected than industrialised countries.

Appendix C

Some groups acting for climate justice

Christian environmental organisations

- Anglican Communion Environmental Network **acen.anglicancommunion.org**
- A Rocha **arocha.org** (see chapter 8)
- CAFOD **cafod.org.uk**
- Catholic Earthcare Australia **catholicearthcare.org.au**
- Christian Aid **caid.org.uk** (see chapter 2)
- Christian Climate Action **christianclimateaction.org** (see chapter 7)
- Climate Stewards **climatestewards.org** (see chapter 4)
- European Christian Environmental Network **ecen.org**
- Green Anglicans **greenanglicans.org**
- Green Christian **greenchristian.org.uk**
- John Ray Initiative **jri.org.uk**
- Lausanne/WEA Creation Care Network **lwccn.com**
- Operation Noah **operationnoah.org**
- Plant with Purpose **plantwithpurpose.org**
- Tearfund **tearfund.org**
- Young Christian Climate Network **yccn.uk**
- Young Evangelicals for Climate Action **yecaction.org**

Secular environmental organisations

- Clean Air Task Force **catf.us**
- Climate Outreach **climateoutreach.org**
- Climate Psychology Alliance **climatepsychologyalliance.org**
- Coalition for Rainforest Nations **rainforestcoalition.org** (see chapter 11)
- Extinction Rebellion **rebellion.earth** (see chapter 11)
- Friends of the Earth **friendsoftheearth.uk**
- Greenpeace **greenpeace.org.uk**
- Project Drawdown **drawdown.org** (see chapter 11)
- Woodland Trust **woodlandtrust.org.uk**
- World Wildlife Fund **worldwildlife.org**

See also the blog at **thinkingaboutcharity.blogspot.com/2020/07/recommended-climate-change-charities.html**

Appendix D

Where to find out more

Throughout this book we have given references to places where you can find out more about the **ecological** crisis and what to do about it. Here we highlight just a few resources that we particularly recommend. For more, see **eco.resilientexpat.co.uk/resources.html**.

Books

Greta Thunberg, *No One Is Too Small to Make a Difference* (Penguin, 2019)
Greta Thunberg's essays (short book which is easy to read).

Ruth Valerio, *Saying Yes To Life* (SPCK, 2020)
More detailed exploration of climate change, including biblical discussion (focusing on the book of Genesis).

Jeremy Williams, *Time to Act* (SPCK, 2020)
Christian Climate Action resource book.

Extinction Rebellion, *This Is Not A Drill* (Penguin, 2019)
Extinction Rebellion handbook.

Robert Henson, *The Rough Guide to Climate Change* (Rough Guides, 2011)
Well-structured and readable overview, good for dipping into.

Videos

Science with Steph
youtube.com/playlist?list=PLRx8CUuVFkNMvGY-inzXI-gtatZZ3LcO9
Short, fun, easy to watch and informative vlogs on the **environment**, recycling and other science topics from a renewable energy specialist. Great for young people.

Christianity and climate change
tearfund.org/campaigns/christianity-and-climate-change-film-series
Tearfund has produced a series of nine short films on Christianity and climate change. Each film lasts for about five minutes and comes with discussion questions and suggested actions. All free to download.

How to live a low-carbon lifestyle
youtube.com/playlist?list=PLlIKCaEd5CsSldUumDIntkfRCNY1EF2NH
Short videos by Tom Bray on different aspects of a low-carbon lifestyle.

(available online)

IPCC report *Global Warming of 1.5°C*
ipcc.ch/sr15
On how we could limit global warming to 1.5C, and what will happen if we reach 2°C. Long and heavily referenced. Try the Summary for Policymakers, and the summary boxes towards the end of each chapter.

David MacKay, *Sustainable Energy Without the Hot Air* (Green Books, 2009)
withouthotair.com
A thorough, practical analysis of how to switch to renewable energy for all the things we need.

Burning Down the House: How the church could lose young people over climate inaction (We Are Tearfund/Youthscape, 2021)
weare.tearfund.org/burning-down-the-house
630 Christian 14–19-year-olds from the UK were asked about their thoughts on the climate, the church and their faith.

Appendix E

Using this book with groups

This book can be used with youth groups, house groups, other small groups and family Bible studies. We suggest reading the Bible passages and discussing the theme of the chapter, the questions and which of the challenges each person plans to try. You could try reading the Bible verses in a different version to see if anything new strikes you – for example, using *The Message*. You might like to watch a short YouTube video together (see Appendix D) and discuss it.

The questions are to help us think – they do not generally have right or wrong answers. We recommend using highlighter pens, or a pencil in the margin, to note sections which are important to you, to remind you when you are in a group discussion.

There are different opinions about the ideas expressed, so allow different views to be discussed in the group.

As a group, can you put your faith, hope and love into action by fundraising together for one of the charities discussed in this book? Group fundraising can be fun and meaningful.

Competitions or challenges can be another fun way to engage members of a group. Possibilities include:

- List all 66 books of the Bible with a page number of where each is mentioned in this book.
- List as many countries as you can which are referred to in this book (with the page number).
- List actions you are taking to help protect the **environment**.

Jamie's glossary

Biodiversity The variety of plant and animal life (it is important to have a lot of variety).

Biofuels Fuels made from waste or **organic** material (like trees or other plants), which therefore produce **greenhouse gas emissions** but can also be replaced over a short timescale (by growing new plants or producing more waste).

Carbon dioxide A gas composed of one carbon and two oxygen atoms. When **fossil fuels** are burned, carbon dioxide is produced. It is a **greenhouse gas**. See Appendix A.

Carbon emissions **Carbon dioxide** entering the air (e.g. produced by transport) that is bad for the **environment**. See Appendix A.

Carbon footprint The amount of **greenhouse gas** emissions a person (or organisation or event) produces through their activities.

Carbon offsetting Paying money to compensate for **carbon emissions** you are responsible for. Examples of carbon offsetting projects include planting or protecting trees, providing wind farms or developing green technologies.

Carbon negative Creating an environmental benefit by removing more **carbon dioxide** from the atmosphere than is produced.

Carbon neutral	Removing as much **carbon dioxide** from the atmosphere as is produced.
Climate justice	A term used to frame climate change as an ethical and political issue, therefore a matter of justice.
Climate refugees	People who are forced to leave their home region due to sudden or long-term changes to their local **environment**.
Drought	When there is no rain for a long time, there is a shortage of water and the land becomes very dry.
Ecological	Things relating to the **environment** or **ecology**.
Ecology	The study of the interaction between living things and their **environment**.
Ecosystem	How plants and animals live together in a particular **habitat**.
Energy mix	The combination of the types of energy sources used in a region. It includes the percentage of different types of **fossil fuels**, renewable energy and nuclear power that contribute to the electricity that is produced.
Environment	All naturally occurring things (living and non-living) and their interaction.
Fossil fuels	Coal, natural gas, petrol, oil, peat and other substances produced by animal and plant material being crushed under rocks for thousands of years, and burnt as fuels. See Appendix A.

Greenhouse gas (emissions) Gases in the air that build up in the atmosphere and trap more heat and make the planet warmer (the 'greenhouse effect'). They include **carbon dioxide**, water vapour, sulphur dioxide, nitrous oxide, ozone and **methane**.

Greenwashing Conveying a false impression or providing misleading information about how an action is better for the **environment** than it really is, or doing one thing to benefit the **environment** in order to try and cancel out action that harms the **environment**.

Habitat A natural home for an organism – for example a rainforest, a desert, a marsh or a coral reef.

Industrial farming Also called industrial agriculture/factory farming, the large-scale, intensive production of animals and crops, often involving harmful use of antibiotics in animals (because of filthy conditions) and chemical fertilizers on crops. It may involve genetically modified crops and practices that increase pollution, deplete the land and mistreat animals. Industrial farming has been linked with causing infectious diseases and pandemics among humans. It is possible to intensify farming sustainably.

IPCC Intergovernmental Panel on Climate Change, set up by the United Nations to provide regular scientific assessments of climate change. Thousands of scientists contribute, and summaries for policymakers are approved by 120 countries.

Methane A **greenhouse gas** composed of one carbon atom and four hydrogen atoms, produced by burning some **fossil fuels** and when animals burp. The same gas as natural gas.

Net zero (emissions)	A point at which the emissions of **greenhouse gases** equal the amount of greenhouse gases captured and stored, so that the net result is zero.
Organic	Made of carbon or living things.
Permafrost	Ground that remains continuously frozen for at least two years.
Sustainable	Able to continue for a long time without damaging the **environment** or using up resources that can't be replaced.

Notes

Preface

1 Sir David Attenborough speaking at COP24, Katowice, Poland, 3 December 2018.

1 Waste

1 lexico.com/definition/prodigal
2 fao.org/3/mb060e/mb060e00.pdf
3 wrap.org.uk/resources/report/food-surplus-and-waste-uk-key-facts
4 unfccc.int/news/un-helps-fashion-industry-shift-to-low-carbon
5 wrap.org.uk
6 loveyourclothes.org.uk
7 freecycle.org is one of many networks available for passing on unwanted goods. Also see ilovefreegle.org/communities, gumtree.com and ebay.co.uk
8 wrap.org.uk/resources/guide/textiles/clothing
9 For instance, some plastic bags have a lower **carbon footprint** than paper or cotton bags, unless we reuse the latter far more than the former. See gov.uk/government/publications/life-cycle-assessment-of-supermarket-carrierbags-a-review-of-the-bags-available-in-2006
10 According to the latest figures at ourworldindata.org/grapher/global-plastics-production
11 bpf.co.uk/Sustainability/Plastics_Recycling.aspx
12 bbc.co.uk/news/world-asia-46518747
13 npr.org/2020/09/11/897692090/how-big-oil-misled-the-public-into-believing-plastic-would-be-recycled
14 blogs.scientificamerican.com/observations/more-recycling-wont-solve-plastic-pollution
15 greenpeace.org/usa/oceans/preventing-plastic-pollution
16 oceanconservancy.org/trash-free-seas/plastics-in-the-ocean
17 uq.edu.au/news/article/2015/09/world's-turtles-face-plastic-deluge-danger
18 learn.tearfund.org/en/resources/policy-reports/no-time-to-waste
19 clipper-teas.com/about/our-sustainability
20 ukgbc.org/climate-change
21 ovoenergy.com/guides/energy-guides/average-room-temperature.html

22 Chapter 7 and Technical Chapter E of **withouthotair.com** show how the most efficient way to use heating depends on our heating system and insulation. For instance, a constant thermostat temperature is better in a well-insulated house (try turning the thermostat down one degree), whereas turning the heating on only when required is better if insulation is poor.
23 According to David Mackay, 'All the energy saved in switching off your charger for one day is used up in *one second* of car-driving. The energy saved in switching off the charger for *one year* is equal to the energy in a single hot bath': **withouthotair.com/c11/page_68.shtml**
24 **globalslaveryindex.org**
25 See **independent.co.uk/news/world/politics/fairtrade-it-really-fair-7717624.html** for a discussion of some of the benefits and problems with Fairtrade.
26 In chapter 9 we discuss eating meat and other food choices.
27 **theguardian.com/global-development/2014/dec/10/climate-change-nicaragua-farming-drought-flood**
28 Downloaded from **germanwatch.org/en/16046**. See page 8.
29 Food past its use-by date may not be safe to eat, but food just past its best-before date usually is. See **food.gov.uk/safety-hygiene/best-before-and-use-by-dates**
30 A typical tumble dryer runs at 2.5kW, equivalent to 225 11W low energy fluorescent light bulbs: **electricalsafetyfirst.org.uk/guidance/safety-around-the-home/home-appliances-ratings**
31 **terracycle.com/en-GB**
32 **fairphone.com**
33 See **usave.co.uk/energy/green-energy-suppliers**
34 For tips, see **cat.org.uk/info-resources/free-information-service/green-living/energy-saving-at-home**
35 **youtube.com/watch?v=roc7xcYtFFU**

2 Want

1 R.A. Emmons, *Thanks! How practicing gratitude can make you happier* (Houghton Mifflin, 2007), pp. 11–12.

3 Reasons to care for creation

1 It does not matter whether you believe the 'days' in the creation story were 24-hour periods, think God created the world over a longer time period or read Genesis 1 as poetry exalting the creator God. The point is that it describes a creator God.

2 C.R. Swindol, *The Tale of the Tardy Oxcart* (W Publishing Group, 1998).
3 **pnas.org/content/early/2020/02/04/1913007117**; **pnas.org/content/117/24/13596**
4 In the Bible, seven is the number of completeness and perfection.
5 **rspb.org.uk**
6 **churchofengland.org/media/19521**
7 **theguardian.com/politics/2003/jul/28/environment.greenpolitics**
8 Extinction Rebellion, *This Is Not A Drill* (Penguin Random House, 2019), pp. 31–32.
9 **politheor.net/the-paradise-paradox-maldives-a-sinking-country**; **ipcc.ch/site/assets/uploads/2018/02/WGIIAR5-Chap29_FINAL.pdf**
10 **today.yougov.com/topics/science/articles-reports/2019/09/16/global-climate-change-poll**
11 **who.int/news-room/fact-sheets/detail/climate-change-and-health**
12 **ipcc.ch/2018/10/08/summary-for-policymakers-of-ipcc-special-report-on-global-warming-of-1-5c-approved-by-governments**
13 **christianaid.org.uk/sites/default/files/2019-07/Hunger-strike-climate-and-food-vulnerability-index.pdf**
14 N. Spencer and R. White, *Christianity, Climate Change and Sustainable Living* (SPCK, 2007), pp. 47–48.
15 **theguardian.com/commentisfree/2014/sep/21/desmond-tutu-climate-change-is-the-global-enemy**
16 In the same way, God will subdue (*kabash*) our sins (Micah 7:19).
17 Spencer and White, *Christianity, Climate Change and Sustainable Living*, p. 84.
18 K. K. Wilkinson, *Between God and Green: How evangelicals are cultivating a middle ground on climate change* (Oxford University Press, 2012).
19 Evangelical leaders: **npr.org/2006/02/08/5194527/evangelical-leaders-urge-action-on-climate-change**; Southern Baptists: **bpnews.net/27585/full-text-of-sbeci-declaration**; Vineyard: **arcworld.org/downloads/Christian-Vineyard-Evangelical-7YP.pdf**; the Pope: **theguardian.com/environment/2019/jun/14/pope-francis-declares-climate-emergency-and-urges-action**; Salvation Army: **salvationarmy.org.uk/id/climate-change**; Methodist Conference: **methodist.org.uk/about-us/news/latest-news/all-news/methodist-church-recognises-a-climate-emergency**; General Synod: **churchofengland.org/media/19521**
20 **margaretbarker.com/Papers/CreationTheology.pdf**
21 Christian Climate Action, *Time to Act* (SPCK, 2020), p. 237.
22 See **whatplastic.co.uk** and **thegreenparent.co.uk/articles/read/make-your-own-cleaning-products**
23 **1millionwomen.com.au/blog/adding-soap-nuts-your-load-pros-and-cons**

4 Love in action

1. C.S. Lewis, *The Four Loves* (Geoffrey Bles, 1960).
2. The hymn 'Lord, you were rich beyond all splendour, yet, for love's sake, became so poor' expresses this beautifully.
3. Over 1,800 people were arrested for their non-violent action against climate change during autumn 2019 in London alone, with many arrests taking place in other countries as well. See **theguardian.com/uk-news/2019/oct/22/extinction-rebellion-protests-cost-met-police-37m-so-far**
4. **noemamag.com/the-ethics-of-the-future**
5. G. Thunberg, *No One Is Too Small to Make a Difference* (Penguin, 2019), pp. 15–16.
6. **voanews.com/east-asia/samoa-pleads-global-action-tackle-climate-change**
7. **globalwitness.org/en/blog/climate-change-front-line-why-marginalized-voices-matter-climate-change-negotiations**

5 Hope in action

1. See video on Deep Adaptation at **youtube.com/watch?v=daRrbSl1yvY**
2. **thecharactercourse.com/session-02-hope**; **chericebock.com/2016/03/12/hope-in-the-hebrew-Bible**
3. **bmsworldmission.org/ecological-hope-in-crisis**
4. S.J. Lopez, *Making Hope Happen: Create the future you want for yourself and others* (Atria Paperback, 2013), p. 24.
5. C.S. Lewis, *Mere Christianity* (HarperSanFrancisco, 2001), pp. 134–35.
6. **chericebock.com/2016/03/12/hope-in-the-hebrew-Bible**
7. **en.wikipedia.org/wiki/Two_Wolves**
8. **iea.org/articles/global-energy-review-co2-emissions-in-2020**
9. **youtube.com/watch?v=OiTrpaPnMYE**
10. **ejfoundation.org/reports/climate-displacement-in-bangladesh**; **ncdo.nl/artikel/climate-change-its-impacts-bangladesh**; **germanwatch.org/en/19777**
11. **bmsworldmission.org/about-us/creation-stewardship-policy**
12. **bmsworldmission.org/wp-content/uploads/woocommerce_uploads/2017/03/Mission-catalyst-3-2014-no-marks-v2.pdf**
13. See **greenmatch.co.uk/blog/2017/04/sustainable-and-smart-cities-around-the-world** and **independent.co.uk/news/uk/home-news/nottingham-carbon-neutral-climate-change-global-warming-emissions-a9287851.html**
14. G. Thunberg, *No One Is Too Small To Make A Difference* (Penguin, 2019), p. 40.

6 Wisdom in action

1 newscientist.com/article/2238831-greta-we-must-fight-the-climate-crisis-and-pandemic-simultaneously
2 Spencer and White, *Christianity, Climate Change and Sustainable Living*, p. 48.
3 theguardian.com/environment/2019/mar/12/air-pollution-deaths-are-double-previous-estimates-finds-research
4 C. Liu, A. Linde and I. Sacks, 'Slow earthquakes triggered by typhoons', *Nature*, 459 (2009), pp. 833–36. See also theguardian.com/world/2016/oct/16/climate-change-triggers-earthquakes-tsunamis-volcanoes
5 Downloaded from germanwatch.org/en/16046
6 See ethicalconsumer.org/money-finance/shopping-guide/current-accounts

7 Travel

1 Some of XR's critics say they chose the wrong countries to protest in. They say Britain is reducing its **carbon emissions** (if we don't count the carbon emitted to make goods imported from overseas) and that they should protest in China and India. It is true that China and India have much greater populations and carbon emissions, even though they are also investing heavily in renewable energy. China tends to imprison or kill protestors. XR is an international movement, and is active across Africa, India and China (Hong Kong) and many other parts of the world.

2 Martin Luther King Jr, 'A time to break the silence', in *A Testament of Hope: The essential writings and speeches of Martin Luther King Jr.*, edited by James M. Washington (HarperCollins, 1986), pp. 231–44.
3 en.wikipedia.org/wiki/Environmental_impact_of_transport; assets.publishing.service.gov.uk/government/uploads/system/uploads/attachment_data/file/790626/2018-provisional-emissions-statistics-report.pdf
4 greencarreports.com/news/1105626_why-motorcycles-may-not-be-greener-than-cars-missing-emission-gear
5 greenchoices.org/green-living/transport/motorcycles
6 transportenvironment.org/what-we-do/electric-cars/how-clean-are-electric-cars
7 youtube.com/watch?v=2b3ttqYDwF0
8 For benefits and problems with **carbon offsetting**, see explainthatstuff.com/carbon-offsets.html
9 europarl.europa.eu/RegData/etudes/BRIE/2018/621893/EPRS_BRI(2018)621893_EN.pdf
10 B. Bilston, *You Took the Last Bus Home* (Unbound, 2016), p. 208.

11 **hovercruiser.org.uk/index.php/home/hovercraft-links/39-enironmental-impact-report**. The chief development officer from HoverAid explains, 'If you can use a boat, use a boat; if you can use a truck, use a truck', but these cannot cross the shallow rivers (and loose sand and mud) where HoverAid often works. 'All communities are our neighbours, and God's love is for all people. If you are going to go to the village on the Mangoky river then a hovercraft is *relatively* a better choice than a Landcruiser, a boat, a light aircraft (no airstrips) or a helicopter (insanely expensive). Ecologically an even better choice is to go by ox-cart or walk, however most of the work we do involves doctors with limited time and we can visit a village in a day that would take two weeks to walk to. The hovercraft makes it possible.'
12 **uk.hoveraid.org/whatwedo/programmes/mms**
13 **rcplondon.ac.uk/projects/outputs/every-breath-we-take-lifelong-impact-air-pollution**

8 A time to plant

1 **sciencemag.org/news/2017/08/trees-amazon-make-their-own-rain**
2 **dec.ny.gov/lands/90720.html**
3 **precisiontreemn.com/tips/14-fun-facts-about-trees.html**
4 **shop.arocha.org**
5 **realclimate.org/index.php/archives/2014/10/how-do-trees-change-the-climate**
6 **rainforestconcern.org/forest-facts/why-are-rainforests-being-destroyed**
7 **wwf.panda.org/our_work/forests/deforestation_fronts2/deforestation_in_the_amazon**; **rainforestconcern.org/forest-facts/why-are-rainforests-being-destroyed**
8 **thinkglobalgreen.org/wetlands**
9 **wwt.org.uk/discover-wetlands/wetlands/benefits-of-wetlands**
10 **rhs.org.uk/garden-inspiration/wildlife/rewild-your-garden**
11 See **ecohome.net/guides/3402/grass-lawn-alternatives-eco-friendly-bee-friendly** for alternatives to traditional lawns.
12 **wragwrap.com/wrapping-paper-facts**; **worcestershire.gov.uk/info/20708/seasonal_waste_reduction_tips/2045/seasonal_waste_reduction_tips**
13 **shop.arocha.org**

9 What shall we eat?

1 **ora.ox.ac.uk/objects/uuid:b0b53649-5e93-4415-bf07-6b0b1227172f**
2 **chathamhouse.org/publication/livestock-climate-change-forgotten-sector-global-public-opinion-meat-and-dairy**

3 nationalgeographic.com/environment/global-warming/methane
4 elementascience.org/articles/10.12952/journal.elementa.000116
5 theguardian.com/food/2018/sep/05/ditch-the-almond-milk-why-everything-you-know-about-sustainable-eating-is-probably-wrong; theguardian.com/environment/2020/jan/07/honeybees-deaths-almonds-hives-aoe
6 sustainablefoodtrust.org/articles/local-alternatives-to-soya-and-palm-meal-will-help-save-rainforests
7 frontiersin.org/articles/10.3389/fsufs.2019.00005/full
8 sciencedirect.com/science/article/pii/S0959378018306101
9 eatforum.org/eat-lancet-commission/eat-lancet-commission-summary-report
10 mashable.com/2015/02/15/food-waste-tips
11 ourworldindata.org/meat-production
12 ourworldindata.org/food-choice-vs-eating-local
13 See robertocazzollagatti.files.wordpress.com/2020/07/certified_palm_oil_habitat_deforestation_cazzollagattivelichevskaya.pdf
14 english.alaraby.co.uk/english/indepth/2019/4/17/climate-change-and-the-yemeni-civil-war; reliefweb.int/report/yemen/desertification-threat-millions-yemenis; awm-pioneers.org/2016/10/yemen-crisis-at-a-glance

10 Mustard seeds

1 Due to the way Esther was looked after by her uncle (Esther 2:10–11), it is thought that she might have been about 14 years old when chosen to be queen, and about 19 when she thwarted the plot to destroy the Jews.
2 G. Thunberg, *No One Is Too Small to Make a Difference* (Penguin, 2019), p. 4.
3 sciencefocus.com/planet-earth/the-thought-experiment-what-is-the-carbon-footprint-of-an-email
4 ovoenergy.com/blog/green/reusable-vs-disposable-nappies-which-is-better-for-the-environment.html; babycentre.co.uk/a559767/choosing-reusable-nappies
5 ethicalsuperstore.com/category/beauty-health-and-wellbeing/menstrual-care/reusable-sanitary-pads
6 nationalgeographic.com/environment/2019/09/how-tampons-pads-became-unsustainable-story-of-plastic
7 simonguillebaud.com/bleeding-beautiful
8 theguardian.com/environment/green-living-blog/2010/aug/19/carbon-footprints-dishwasher-washing-up; blueland.com/articles/what-is-the-most-sustainable-way-to-wash-your-dishes
9 M. Berners-Lee, *How Bad Are Bananas?* (Profile Books, 2010).
10 ukscn.org/the-green-new-deal

11 ukscn.org/our-demands
12 un.org/en/chronicle/article/climate-change-and-malaria-complex-relationship; dailytimes.com.pk/104158/climate-change-and-vector-borne-diseases-in-pakistan; doi.org/10.1371/journal.pone.0224813; carbonbrief.org/ebola-epidemics-will-increase-with-greenhouse-gas-concentrations-study-finds; who.int/news-room/q-a-detail/q-a-on-climate-change-and-COVID-19; unenvironment.org/resources/report/preventing-future-zoonotic-disease-outbreaks-protecting-environment-animals-and
13 reshareworthy.com/fire-caused-by-phone-charger
14 theguardian.com/environment/2006/sep/23/water
15 See also statista.com/statistics/827278/liters-per-day-household-water-usage-united-kingdom-uk and projectsolaruk.com/blog/5-astonishing-facts-carbon-footprint

11 Moving mountains

1 un.org/sustainabledevelopment/sustainable-development-goals
2 ourworldindata.org/co2-and-other-greenhouse-gas-emissions#emissions-by-sector
3 iopscience.iop.org/article/10.1088/1748-9326/aa7541
4 youtube.com/watch?v=E0i5OnJnUkA
5 drawdown.org
6 medium.com/@tsloane/top-charities-for-climate-change-54ddf6911e1
7 drawdown.org/solutions/health-and-education
8 withouthotair.com/Contents.html
9 pubmed.ncbi.nlm.nih.gov/18258860
10 withouthotair.com/c24/page_161.shtml
11 See youtube.com/watch?v=QdKdCqZQpA0, catf.us/work and drawdown.org/sectors/electricity
12 founderspledge.com/stories/climate-change-executive-summary
13 ipcc.ch/sr15
14 bbc.co.uk/news/science-environment-46455844
15 theyworkforyou.com
16 fullfact.org/news/are-100-companies-causing-71-carbon-emissions
17 bp.com/en/global/corporate/news-and-insights/press-releases/bernard-looney-announces-new-ambition-for-bp.html
18 concepts.effectivealtruism.org/concepts/divestment; theconversation.com/fossil-fuel-divestment-will-increase-carbon-emissions-not-lower-them-heres-why-126392; worldbank.org/en/programs/pricing-carbon
19 rebellion.earth
20 climateassembly.uk/report
21 See youtube.com/watch?v=qTgMTeGqKv8

22 See W. Wink, *The Powers That Be: Theology for a new millennium* (Bantam Doubleday Dell, 2020).
23 **founderspledge.com/stories/climate-change-executive-summary**
24 This was because developing countries like Brazil said it would be unfair on them. Developed countries like Britain had become rich partly by destroying rainforests in Africa, Asia, and South America. Now developed countries were saying developing countries could not develop. It was like buying all the best games and then closing the website so no one else could buy them!
25 **akgeo.org/youth-programs/arctic-youth-ambassadors-2017-2019**
26 Search for 'sustainable countries' at **buzzonearth.com**

12 Join the dream team

1 Some argue that this is not in Wesley's writing. Whoever said it, it is worth taking note.

Appendix A

1 Adapted from **data.giss.nasa.gov/gistemp/graphs_v4/graph_data/Global_Mean_Estimates_based_on_Land_and_Ocean_Data/graph.html**
2 **jpl.nasa.gov/news/news.php?feature=7159**
3 **climate.nasa.gov/evidence**
4 **gov.uk/government/publications/thames-estuary-2100-te2100/thames-estuary-2100-te2100**
5 **smithsonianmag.com/innovation/is-a-lack-of-water-to-blame-for-the-conflict-in-syria-72513729**
6 **visionofhumanity.org/wp-content/uploads/2020/10/ETR_2020_web-1.pdf**
7 **skepticalscience.com/Global-Warming-in-a-Nutshell.html**
8 Adapted from **ourworldindata.org/atmospheric-concentrations**, combining data from **ncdc.noaa.gov/paleo-search/study/17975**, **scrippsco2.ucsd.edu** and **esrl.noaa.gov/gmd/ccgg/trends/data.html**. Reproduced under Creative Commons licence CC-BY.
9 R. Valerio, *Saying Yes To Life* (SPCK, 2020), p. 128.
10 **cbd.int/gbo/gbo5/publication/gbo-5-en.pdf**
11 **livingplanet.panda.org/en-gb**
12 **sciencedirect.com/science/article/pii/S0006320718313636**
13 **britannica.com/story/what-would-happen-if-all-the-bees-died**
14 **public.wmo.int/en/resources/united_in_science**
15 **exploratorium.edu/climate/ice**
16 **royalsocietypublishing.org/doi/10.1098/rsta.2010.0292**

Appendix B

1. ec.europa.eu/jrc/en/publication/fossil-co2-emissions-all-world-countries-2018-report
2. christianaid.org.uk/sites/default/files/2019-07/Hunger-strike-climate-and-food-vulnerability-index.pdf
3. See planetforward.org/idea/bhutanese-carbon-neutrality; climateactiontracker.org/countries/bhutan; gviusa.com/blog/bhutan-carbon-negative-country-world
4. Downloaded from germanwatch.org/en/19777

About the authors

Dr Debbie Hawker is a clinical psychologist who supports mission workers around the world. She treats people after trauma, but prefers to *prevent* trauma when possible (including from the catastrophic impacts of climate change). Debbie has written eight previous books. She has not driven a car for over 20 years, does not use a smartphone and can't remember buying herself any new shoes or clothes in the past ten years.

Dr David Hawker is a clinical psychologist, but has also been known as the cycling psychologist, because he cycles wherever he can. He specialises in working with mission workers' children and families. He is also a safeguarding administrator and believes that one of the biggest risks to young people is the impact of climate change. He likes to eat up leftover food so that it is not wasted. At university, a tutor told David he didn't need to write tiny notes in old exercise books just to save paper.

Jamie Hawker is 14 years old (as of 2021) and is considering becoming a train driver, ecologist or cartographer, among other jobs. He has enjoyed train travel for many years. He also enjoys geography. He is a Green Agent of Change with the Methodist Church.

Resilience in
Life and Faith
Finding your strength in God
Tony Horsfall and Debbie Hawker

MARTIN J. HODSON &
MARGOT R. HODSON
A CHRISTIAN GUIDE TO
ENVIRONMENTAL
ISSUES
FOREWORD BY
ANDY ATKINS
CEO, A ROCHA UK
FULLY REVISED AND UPDATED SECOND EDITION